NEW TECHNOLOGY AND WORK IN THE FUTURE

NEW TECHNOLOGY AND WORK IN THE FUTURE

Edited by

TONY CHARLES

GIAN PUBLISHING HOUSE

NEW DELHI - 110002

GIAN PUBLISHING HOUSE
4-C Ansari Road, Daryaganj
New Delhi 110002

ISBN—81—212—0283—3

PRINTED IN INDIA

Published by Mrs. Gayatri Garg for Gian Publishing House, New Delhi and Printed at Godwall Printers, Delhi.

Preface

The papers which comprise this volume were presented at the XI World Congress of Sociology, New Delhi, August 1986. The International Sociological Association contains thirty-eight research committees, and of these, the Futures Research Committee provided the forum for these papers. Futures research is global in its perspectives—it raises fundamental questions about the direction of social change, current trends and likely scenarios for all aspects of world society in the future. The Futures Research Committee provides an important network for the exchange of ideas between sociologists and social scientists from all parts of the world.

Discussion of social futures must inevitably take account of the social consequences of new technology and the various ways in which technology impinges upon our lives. This theme was chosen for the sessions of the Futures Research Committee at the World Congress and the work included in this volume represents a continuing process of on-going research into new technology. The Congress sessions in this area included : the social consequences of new technology for work, private life, culture, public services, and developing countries.

The papers in this volume concentrate upon the consequences of new technology for work and paid employment. 'New Technology' in this context is defined as information technology, although there is also a discussion of more traditional forms of manufactutre—a recognition of the fact that 'new technology' is a culturally relative concept. However, the human issues surrounding technology have always raised similar fundamental questions in the social sciences : questions of power and control, public accountability, social values and attitudes, employment and unemployment, the quality of work, the relationship between work, private consumption and leisure, the division between public and private life, and the nature of social innovation.

Several of these issues are discussed in these papers and despite a diversity of approaches and subject-matter, there is a common theme—changes in the nature of work and the future with new technologies is a social issue involving human choices. The conception of technological change as a purely technical or engineering problem in which only questions of 'efficiency' are addressed conceals the real issues confronting the use of technology at the workplace and in the wider society. The choices with new technology concern every aspect from innovation and design through to implementation and they reflect the wider social factors which impinge upon the world of work ; class, gender and ethnic inequalities, the role of trade unions, management and consumers as well as global relations between nation states.

The various papers in this volume do not propose policy solutions but present the social issues involved with new technology and report the findings of extensive sociological research with a bias towards the West European experience. In this way it is hoped that the contents will appeal both to social scientists and students and to a wider audience of policy makers, managers, engineers and scientists. Finally, we hope there is something of interest in the book for the general reader who may already be experiencing the impact of technology at his or her place of work or indirectly through changes in culture and patterns of consumer spending.

The choice of India as the venue for these papers was wholly appropriate : for the mix of participants to be truly international and because India has, for so long, been a country which stimulates the sociological imagination. All the contributors to this volume would like to put on record their thanks to our Indian colleagues who worked to make the

Congress a success and to all those participants at the conference sessions who informed debate on the papers but could not be recorded here.

TONY CHARLES

Head
Dept. of Sociology
Staffordshire Polytechnic
Stoke-on-Trent ST 4 2 DF (U.K.)

Introduction

The papers in this collection represent a broad spectrum of current sociological research into new technology and work which has focused primarily on the advanced industrial societies. With the exception of Kumar's paper on India, which is an important corrective to the exclusive focus upon technology in the 'First World', the various contributions introduce the reader to one of the major areas of sociological research in Western Europe and the USA, which is a reflection of the revolutionary changes which are underway in the nature of work and employment in the context of developments in micro-electronics and telecommunications. At the same time it is clear that these changes are occurring in a global context in terms of the international economy and division of labour such that the social consequences of new technology are not confined to the West but will extend to the Third World.

Much of the research reported here concentrates upon changes in the quality of work, employment and organisational change at the enterprise or industry level in industrial societies. Decisions about the use of new technology are taking place in the context of world economic restructuring, decline in traditional manufacturing in Western economies, more intense international competition and more sophisticated markets for products which are changing the 'rules of the game' in terms of markets for products, services and labour. Strategies of employers and the reaction of organised labour reflect these changing conditions, particularly the move towards 'flexible specialisation' and the search for 'niches' in the market in contrast to traditional mass-production based upon standardisation and price competition. The contributions which follow contain a number of possible scenarios for work and employment in the future in the West, but what are the possible implications for the Third World, and India in particular ?

New technologies entail both opportunities and dangers ; on the one hand, new types of skill emerge along with possibilities for more flexible and participative organisational forms, but, on the other hand, there is ample evidence of job deskilling, polarisation of the workforce and widespread labour displacement. 'Winners' and 'losers' in this process of change can therefore be identified in Western societies. Similarly, consequences for Third World countries are likely to exhibit some positive aspects but also a number of negative features which could increase inequalities between rich and poor nations. There is no technological determinism involved and the assumptions, of earlier neo-evolutionary theories of 'technological lag' are clearly inappropriate for understanding the nature of interdependence between industrial nations and developing countries. In the same way that the social consequences of new technology for work in industrial societies reveal significant variations, possible implications for developing countries are not likely to be uniform.

In S.E. Asia, where there has been a shift in manufacture and final assembly from the advanced industrial societies there is the possibility that manufacture may return in part to the West as the advantages of cheap labour recede in the face of highly automated production. 'Enclaves' of high technology industry in the 'free trade zones' of newly industrialising societies may, therefore, be vulnerable depending upon decisions of trans-national corporations. Secondly, there is the possibility that the gap between rich and poor countries will widen since the advantages of new technology will accrue firstly to the advanced industrial societies. Changes in the electronics and garment industries provide clear examples of likely trends. In both these sectors, direct labour is becoming less important. In the garment industry, computerisation of cloth cutting, pattern design and assembly is changing the character of a traditionally labour intensive sector of employment.

At the same time certain Third World countries, including India, have developed a significant industrial base, and are themselves the home of trans-national corporations. Based upon policies of import substitution, the availability of skilled graduate manpower (in computer software, for example) and political control over capital inflow and outflow, these countries may enter the 'race' or develop their own protected industries for mass-produced products. Other Third World countries will be out of the 'race' altogether, and the effects of new technology will be distant and problematic especially where technological transfer is based upon decisions of trans-national corporations.

Changes in the division of labour in the West may also be reproduced in certain Third World countries—in services, for example, computerisation in banking, airlines, and government departments may increase the dislocation and polarisation of the workforce with a core of qualified workers in this new service sector and a mass of unemployed or under-employed labour in the more traditional sectors.

However, the focus of these papers is one step removed from the preceding discussion of international implications of technology transfer. The papers variously concentrate on the ways in which new technology is being implemented in the workplace and the reactions of the people most affected : on the shop-floor or in the office. The contributors have tried to capture the changes at work for a wide variety of workers in different sectors, from manufacturing to new service sectors and from traditional manual workers to clerks and professionals. The concept of work used in the papers refers primarily to paid employment but some of the papers also reflect a growing awareness in sociological analysis that 'work' also includes much more than employment—work extends into the private household in the form of self-provisioning and domestic tasks which are unpaid. Technological changes also permeate consumer behaviour and private life as well as the culture of modernity. Work in this volume is therefore treated in the widest sense of the term to acknowledge this interdependence between work and other features of life in modern society.

In the first paper, I develop a selective over-view of current research into new technology and work. There is a discussion of the major theoretical approaches in sociology concerned primarily with the relationship between technology, waged work and the division of labour in capitalist societies. This is followed by an examination of the most recent empirical research in manufacturing and the service sectors which has focused upon changes in the quality of work for both manual and non-manual workers. There is also an extensive bibliography to this paper to enable the reader to follow up key issues identified in the literature. This is followed by a second paper from the Editor which concentrates in more detail on changes in clerical work. Here, the theoretical debates are complemented by case studies and empirical evidence from the financial services sector and public sector health provision.

Suresh Kumar presents the results of a case study in India of a more traditional form of manufacturing technology. The effects of a new steel plant on the local community are examined in terms of changes in occupational structure and traditional rural ways of life. The tension between traditional culture and new forms of work illustrates many of the fundamental concerns which have always been central to sociology.

The papers by Werner Wobbe and Peter M.Van Der STAAL are concerned with new technology in manufacturing. Werner Wobbe outlines the research programme into new technology sponsored by the European Economic Community and identifies the major issues facing European countries in terms of future trends in work and employment. The FAST (Forecasting and Assessment in Science and Technology) programme of the EEC represents the growing importance of co-ordinated cross-national comparative research in the social sciences and the considerable output of this research initiative is documented for the reader to select from the range publications. Peter M. Van Der STAAL examines new technology applications in small and medium sized enterprises which are crucial to the European economy. Using the Netherlands as a case study, the impact of automation is discussed in manufacturing and services in terms of the nature of technological diffusion and the significance of fiexible manufacturing and computerised information systems which are now developing in smaller firms compared to the earlier forms of automation which were largely confined to large enterprises and mass-production.

Three of the contributors focus the implications of new technology for non-manual work. Moritz Rosenmund develops the initial hypotheses which are guiding his empirical research into the professions who work with the new information technologies. New forms of the division of labour are identified based upon changes in the total information process in organisations which challenge traditional specialist identities and prevailing images of work. Juliet Webster's paper provides data from case studies of office automation and word processing. The consequences for the work of female office workers did nct follow a uniform pattern and they reflected prior forms of the division of labour rather than intrinsic technical features of computers or word processors.

Denise Deliege presents the results of research into the uses of imformation technology in health services as a major example of the widespread diffusion of new technology for non marketed services. Focussing on health provision, her paper applies hypotheses derived from Toffler's *Third Wave*, in order to analyse possible consequences for health care professionals and patients. Compared to some of the other papers, a relatively more optimistic scenario for changes in work and for the consumer is presented in which information technology allows the possibility for more participative forms of health care and greater access to information for medical decision making by professionals.

The final three chapters contain contributions which widen the debate about new technology beyond the work-place and build upon much of the research in the earlier chapter to present longer term scenarios for the future. Anna Coen presents a strategic long term policy plan using her concept of the 'social multiplier effect'. New social values are identified which shift from quantity in work and life. The economic emphasis on GNP (Gross National Product) as an indicator of growth is reduced in Coen's scenario of the rise of more proactive lifestyles which challenge the dominance of both market mechanisms and the state. The final paper by Victor Scardigli identifies the choices and dilemmas presented by the possible futures with the 'digital society'. The relation between technique and culture can take three forms : harmonious integration, conflict and rejection or dependency and subjection. All three possibilities are identified in the paper with their corresponding opportunities and threats both for the culture of everyday life and the wider inrerrelations between rich and poor countries.

Together, the collection of papers draw together much of the current international sociological research into new technology which touch people as workers and as consumers. The nineteenth century belief in technological progress based on evolutionary assumptions is rejected in these papers. Instead, the dilemmas and choices for the future with new technology are emphasised. Above all, the papers aim to stimulate debate about choices for new technology and they contribute a sociological dimension to the analysis of technology and culture as a corrective to the dominance of economic and technical criteria for technology policy. They will have achieved their objective if they encourage a critical discourse which allows for the social shaping and human control of technology rather than the technical determination of society and culture in the future.

Contents

Contributors

Tony Charles, Acting Head of Department of Sociology, Staffordshire Polytechnic, UK. He was convenor of the sessions on work and new technology at the World Congress of Sociology, New Delhi, 1986. He took a degree in Sociology and Philosophy at Southampton University (1968) and subsequently undertook postgraduate studies and research at Leicester University (MA) and the London School of Economics. He has published and researched on the social implications of new technology. He spent a year researching new technology in Swiss Banks and has been associated with comparative technology research sponsored by the European Community. He is currently contributing to a new dictionary of Sociology to be published in 1990.

Suresh Kumar, Reader in Sociology at Ravi Shankar University, Raipur, India. He has a degree in Philosophy, an MA in Sociology (Sanskrit and traditional culture) and a Ph. D. He is the author of *Social Mobility in Industrializing Society* and a number of articles. He has been a consultant to a research project at Bhilai Steel Plant. He is also a member of the Indian Sociological Society.

Werner Wobbe, Research officer at the European Commission in Brussels. He is responsible for the Technology, Work and Employment Research

Programme at FAST (Forecasting and Assessment in Science and Technology). He has a Ph. D. from the University of Gottingen, West Germany where he was a research fellow before taking up a post at the EEC. He is an expert on the Sociology of new technology in manufacturing and is the author of two books on *Robots and Work in the Car Industry* and *The Factory of the Future*.

Peter van der Staal, Professor of Futures Research at the University of Leiden in Holland. He has degrees in electronic engineering and in the methodology of the social sciences. He has worked and published in technology forecasting and assessment and industrial sociology at Delft University of Technology, Holland.

Moritz Rosenmund, holds a research post in the Department of Sociology, University of Zurich in Switzerland. He has researched and published in the general area of the social impact of new technology. He was Co-Chairman of the Sessions on New Technology and Work in the Future at the World Congress of Sociology, New Delhi.

Juliett Webster, Research Fellow at the University of Edinburgh where she is working on a major government funded research programme—PICT, Programme for Information and Communications Technologies. She holds a degree and Ph. D. from the University of Bradford, U.K. She has published a book on Office Automation and a number of articles. She was previously a Research Fellow at the City University in London.

Denise Deliege, Professor of Social Sciences at the University of Louvain in Belgium and consultant for the World Health Organisation. She is Head of the Department of Health Services Research. She has a B.A. degree in Philosophy and a Ph. D. in Sociology. She has published widely in the field of demography, economics and sociology of Health. She is now researching into medical manpower planning.

Anna Coen, Researcher for ISPE (Instituto Studi Programmazione Economica), Rome, Italy. She is Vice-President of the Futures Research Committee of the International Sociological Association and a member of the World Futures Studies Federation. Her special interest is futures research, long-term development scenarios and changes in lifestyle where she has published widely. She was Chairperson of Sessions on the Social Consequences of new Technology and Public Services at the World Congress of Sociology, New Delhi.

Victor Scardigli, Research Fellow at CNRS, Centre National De La Recherche Scientifique, Paris. He has researched and published on the relation between technology, culture and ways of life. He is co-author of *La Societe Digitale*, 1984.

1

New Technology and the Future of Work

Tony Charles*

1. Introduction

New technology and work is now one of the largest areas of social research with a virtual publication explosion over the past ten years. What follows is not therefore a comprehensive review of the literature which would be impossible in the space of a short paper. Instead I propose to focus upon significant theoretical developments and differences of approach in the area, together with a discussion of the state of empirical findings. The latter has had to be even more selective and discussion is confined largely to European empirical research, although comparable examples could easily have been cited from N. America, Japan (Littler, 1982) or even the Eastern block countries (Grootings, p. 1986).

At the outset, however, some clarification of the terms 'new technology' and 'work' is required since they are by no means self-evident, in fact they are themselves the subject of theoretical dispute—see, for example, Child, 1986, on the concept of technology and Pahl, 1985, on definitions of work. 'New technology' is clearly both culturally and

*Dept. of Sociology, Staffordshire Polytechnic, U.K.

temporally relative; in this context I am referring to the hardware of microelectronics and the software of systems design applied to :

(a) Manufacturing—which includes, computer—numerical —controlled machines, robots, flexible manufacturing systems and computer-aided design.

(b) Services—computerised information and filing systems, computerised payment systems (in, for example, banking and retail services) which are increasingly combined with new telecommunications technology.

(c) Offices—mainly word-processing and electronic mail.

The above is simply a list of applications which had their origins in early valve computers and the post 1945 development of electronics. The 'new technology' of the 1990s could well prove to be a combination of microelectronics with the technologies of light and also 'biotechnology', but to venture predictions about these developments would be highly speculative at this time. A sociological analysis of technology, however, clearly involves more than reference to new machinery or applications; we are talking about the ways in which material artefacts, machines are inter-related with forms of knowledge and skills and a given technical division of labour.

Thus, the sociology of technical change has attempted to develop classifications of technological levels and analyse changes in the technical division of labour. Well known typologies were developed in the 1960s, describing the change from early mechanisation to assembly-lines and full automation (Blauner, 1964; Woodward, 1965). These are now dated since continuous process industries were taken as characteristic of the automation stage, but applications listed above are now being used to progressively automate batch production. Bright's (1956) classification of 17 levels of automation was important because it focused upon the nature of technology as a social phenomenon via an analysis of changes in man-machine relationships and the progressive substitution of machines for manual and then mental skills. More recent research has moved away from unilinear evolutionary typologies of automation in order to make sense of current trends in advanced manufacturing systems. The distinction between primary, secondary and tertiary mechanisation, for example, refers to different dimensions of automation respectively, transformation of work pieces, transfer and control (Bell, 1972). Coombs (1984) has further analysed these phases of automation in which recent developments in manufacture point to a form of 'flexible tertiary mechanisation' (Coombs : 156, 1984) which can be situated in long waves of economic growth. These typologies do not include technological change in the service sector, but we could note as a general rule point that compared to manufacturing, office work has traditionally not been as highly mechanised ; however, this is currently the sector where information technology may have the greatest effect in terms of future employment and the nature of work with a concentration upon process innovation as much as product innovation (Gershuny, 1986). We shall return to the analysis of technology later since the utility of various classifications is best assessed in relation to the major theories concerning changes in work and the issue of whether or not technology is 'neutral' in terms of its consequences for work organisation.

Turning to a clarification of the concept 'work' in this paper, most of the research to be discussed refers to paid employment. The discussion is also limited mainly to the quality of work and the organisation and design of jobs. Despite futuristic scenarios of a leisure society and the collapse of conventional forms of work, waged-work will remain of central importance at least until the end of this century. Nevertheless, the significance of new technology for unpaid work, the informal economy, households and consumption patterns is now recognised in an expanding area of social research, some of which is represented elsewhere in this volume (see also the review by Gallie, 1985). Also, despite the focus upon the content of work, quantitative and qualitative analysis of unemployment trends cannot be ignored nor divorced from research into the quality of work. For example, changes in the labour market clearly affect employees bargaining position in relation to the introduction of new technology; likewise the displacement of certain categories of workers by new technology can be regarded as the logical end result of a de-skilling process.

The rest of this paper comprises an examination of three main perspectives on new technology and work, followed by more detailed discussion of manufacturing, services and office work.

2. Three perspectives on work and technology

The first of these perspectives is derived from the theory of post industrial society. The logic of technological development implies a growth in knowledge, work, demands for new skills and hence an upgrading in general skill levels. Generally, this approach, based upon a positivistic view of natural science, assumes that technology is outside of politics and ideologically neutral. Clearly this perspective has a long history in Sociology, going back to Durkheim's analysis of the division of labour, convergence theory in the 1960s (Kerr, 1970) and the works of Bell, (1974) and Galbraith, (1967). Subsequently, Wilkinson, (1983) has traced the development of what he calls the innovation approach to new technology with reference to examples from industrial sociology from the 1960's onwards. It is also possible to locate the theory of human capital within the broad assumptions of this perspective (Penn, and Scattergood, 1985). In short, the various approaches within this model emphasise technology as an independent variable in the process of social change in which the direction of change is towards skill upgrading and increase in education and training for the workforce.

The direct antithesis of the first perspective is found in labour process theory. Recent theoretical debate on new technology and work has been dominated in sociology (Anglo-Saxon countries particularly) by discussion of the labour process and the argument that the drive to capitalist accumulation and control over the labour process results in forms of technology which de-skill labour, whether manual or non-manual (Bravarman, 1974, Zimbalist, 1979). It is unnecessary in the space of this paper to review the ten years of subsequent debate and critique connected with the revival of the Marxist theory of the labour process; (see, for example: Wood, 1982; Thompson, 1984 for discussion of much of the literature). The evidence for de-skilling will be discussed in section 3 below, but the theoretical contrasts with the post-industrial perspective are clear : both theories in their original form were deterministic, positing either global de-skilling or upgrading respectively but the conceptualisation of technology was different in each case. For post industrial theory technology was an autonomous development of scientific knowledge in which the technology itself was deemed to have 'effects' on the division of labour. Labour process theory, on the other hand, was not technologically determinist, indeed, for Bravarman (1974) de-skilling was a result of management and Taylorism in particular; mechanisation was subordinate to the requirements of capital and 'technical efficiency' for control over the labour process. Bravarman's work, along with subsequent studies such as Noble's (1979) analysis of the introduction of NC machine tools in the USA were influential in discrediting the view of technology as a neutral force of production.

However, subsequent debate has led to three important lines of qualification in relation to the original Bravarman thesis : the existence of alternative strategies of managerial control (Friedman, 1977); the importance of product and labour markets in terms of their effect upon both management objectives in implementing new technology and worker's capacity to resist policies of job degradation (Edwards, 1979); and finally evidence of job upgrading and re-skilling which leads us to the third perspective on work and technology which can be identified under the umbrella term 'contingency theory'. Rather than denoting a unitary theoretical perspective, I shall use this term as a convenient label to include both the particular perspective within organisation theory (Lawrence and Lorsch, 1967) and its offshoot 'strategic contingency theory' (Child, 1972) as well as a variety of surveys and case studies which treat technology as one among a number of contingent factors affecting changes in the nature of work.

These studies have in common an emphasis upon empirical

research in order to examine the significance of variations in the quality of work at the enterprise, employment sector or national levels despite the introduction of broadly similar forms of new technology. Thus, variations in job design have been found which exhibit both strategies of de-skilling and of polyvalence or multi-skilling (Child, 1985) depending upon the enterprise in question, the market context, culture, and the power of different occupational groups to influence organisational change and the design of work. Emphasis is placed upon the flexibility of electronic technology compared with earlier forms of mechanisation (Child, 1985 ; Coombs, 1985) and therefore the outcome of technological change is seen as indeterminate and cannot be deduced a priori.

Whilst this categorisation of theoretical perspectives may help in understanding the broad trends of development in research, there is always the danger of over-simplification. For the sake of clarity it should be noted that writers within the Marxist labour process perspective have also been critical of the original Bravarman thesis ; the possibility of re-skilling, for example, is not excluded by labour process arguments which see this as part of management strategies to divide and rule or maintain a minimum of workforce consent and co-operation where policies of Taylorism have been counter productive or ineffective as a form of control (Littler, 1982; Buroway, 1979; Friedman, 1977).

The relative merits and limitations of the above perspectives will be discussed in relation to recent research into new technology in manufacturing and the service sectors and in particular, ambiguities surrounding the concept of skill and the problems of comparability of research findings.

3. New technology and manufacturing

There has long been a debate about the relation between technology and work organisation in industrial sociology which focused in the 1960s and 1970s on alienation and worker's attitudes to technical change. Studies of manufacturing of this genre made reference particularly to car assembly and later continuous process in chemicals and oil refining (see for example: Blauner, 1964; Goldthorpe, 1968; Touraine, 1962; Gallie, 1978). The picture of manufacturing which emerged depended largely upon the type of production process—high alienation and job degradation symptomatic of mass production; the persistance of craft skills in small batch production and re-integration of work combined with increased control with automation in continuous process production. There is more or less universal agreement about the job degradation effects of Fordist mass production, it is the analysis of automation which has generated most controversy. We therefore need to look at more recent changes in manufacturing technology mentioned in the introduction to see how the picture of manufacturing industry is changing

Developments since the 1970s in computer aided design (CAD), computer numerically controlled machine tools (CNC), direct numerical control (DNC), and now flexible manufacturing systems (FMS) involve a much wider diffusion of automation to all aspects of the production process. Furthermore, these changes are being implemented in small batch production and in small/medium sized companies in contrast to the post-war systems of mass production. The significance of these developments for work organisation is only now beginning to be investigated in empirical research.

Studies of CNC (machine tools which have a mini-computer on the machine itself programmed to control operations) indicate considerable variation in new work arrangements which are clearly not the result of any technical logic. Emphasis on the flexibility and choices available for work design and skill utilisation is apparent in, for example, the cross-national comparison between Britain and W. Germany by Sorge, *et al* (1983); in the latter there was a trend towards the integration of programming and operator skills on the shop floor thus building upon craft skills, particularly in small-batch production. In Britain, programming and operator functions were more likely to be separated, reinforcing the distinction between white-collar and shop floor levels and greater skill polarisation. These differences are explained by the contrasting traditions of training and apprenticeship in the two countries with the greater emphasis on skilled labour in West Germany, which reflect wider cultural and institutional differences. The case studies by Wilkinson (1983) in Britain reveal important variations between enterprises in the same country. Evidence of skill polarisation and operator de-skilling was apparent in a number of the case studies with the separation of programming and maintenance from operators, but a minority of cases revealed a shift to polyvalence among operators who officially or informally corrected faults in programmes or performed maintenance functions.

The most recent developments in manufacturing involve increased computer integration of processes compared to isolated CNC machines : such as the integration of computer aided design with computer-aided manufacture and FMS. The latter integrates previously discrete elements of production whereby several CNC machines and transfer are directed by a computerised control system so that various parts can be processed in re-programmable sequences. The consequences of these technologies combined with radically different market conditions for products in advanced capitalist societies signal, it is argued, a possible revolution in work organisation (Sabel, 1982, Kern & Schumann, 1984). Market conditions in the present recession are characterised by rapid changes in product demand, intense competition, high product differentiation and away from standardised mass production. These trends have been recognised in the attempt to link 'long-wave' macro-economic theory to managerial strategies in implementing new technology (e.g. Coombs, 1985). The result is a search for greater flexibility in production compared with the 'Fordist' strategy of cutting production costs for standardised goods (Sabel, 1982).

For Kern and Schumann (1984) these changes herald the emergence of a 'new production concept' which reverses earlier Tayloristic forms of work organisation. The logic whereby machines replace firstly physical and ultimately mental human skills is halted for certain forms of production. Integration of tasks is therefore substituted for job fragmentation and hierarchical division of labour to make full use of the flexibility potential of the new technology. Evidence for this trend is derived from research into the machine tool industry, chemicals, and car production in West Germany ; the authors revise their earlier influential thesis of polarisation (1973) which posited a division between a minority of highly qualified jobs and residual unskilled jobs that had not been automated (the similarities with the Bravarman thesis are interesting but the W. German research did not support the thesis of general de-skilling).

Recent European comparative research also identified similar trends although marked variations were evident in terms of changes in skill and the division of labour at both the enterprise and industry levels (Charles, FAST, Turin, 1986). Much depends on prior skill characteristics of the workforce, labour markets and the pre-existing technical division of labour, thus average skill levels in metal machining are high and new activities based upon maintenance, programming and monitoring may be added to the work of the skilled operator where increased integration of production processes creates new forms of work complexity involving a knowledge of the whole system. However, polyvalence is not fully implemented and the continuance of work fragmentation and specialised job demarcations between programming, maintenance, and operators persist.

The optimism about 'new production concepts' is not shared by all researchers (Briefs, 1980), nor by trade unions such as IG METAL (Blum, 1984 & 1986) who have conducted their own research in West Germany. Firstly, the consequences for unemployment are already apparent. It is unlikely that CAM or FMS will lead to the 'unmanned factory' in the foreseeable future but the need for labour will be much smaller due to the integration of tasks, automation of unskilled work, increased shift working (where one shift may be virtually unmanned) and a general intensification of work. In the IG Metal survey (1986) of the social effects of NC/CNC machines, worker reaction was negative for a majority of respondents with reference to mental stress, working speeds and work variety. Similarly, investigations into the use of industrial robots in car manufacture (Wobbe, 1982) revealed increased control by the machine cycle, social isolation and the creation of 'Restarbeitsplatze' or left-over de-skilled tasks such as pallet loading. Similarly, case studies of FMS in the USA (Fadem, 1984) show that Tayloristic criteria still underpin work design such that human elements are removed as far as possible in the new systems. Similar points have been made by Cooley (1982) with reference to the underlying philosophy of engineering design : technicist criteria predominate and systems are machine rather than human oriented.

Questions about design philosophy are also clearly related to the issue of participation and the role of trade unions in the introduction of new technology. Despite the growth in new technology agreements trade unions in Europe have rarely been involved in the early planning stages of new technology and negotiation has tended to be reactive—about manning, pay, conditions of work, rather than the design of technology itself (Martino, 1986; Williams, 1986).

At the time of writing the extent of FMS technology in actual operation is low, with an estimated 25 units in the EEC of which the majority are in West Germany (EEC Commission 1986). 'Flexible manufacturing cells' are more widespread, but less dramatic in their effect since they constitute islands of advanced automation which exist alongside traditional forms of technology. The robot population is higher with over 5000 in use in West Germany, and about 2000 each in France and the U.K. Trends are therefore difficult to discern, particularly about the ways in which FMS is being used to re-structure work but it is possible to conclude this section with some general observations on the findings which are relevant to the earlier discussion of theoretical perspectives.

Firstly, labour displacement at the enterprise level is potentially high, maintenance and control work increases in importance and the number of direct operatives decline in absolute terms and in proportion to other workers. There is some evidence that skilled workers tasks may be enlarged to integrate and include planning and control functions ; a form of polyvalency which could be described as flexible specialisation. This is more likely to occur where the prior technical division of labour was based upon a high skill profile of workers and market conditions put a premium on flexibility and quality rather than mass production. At the same time proponents of the new production concept as well as critics, qualify the thesis of polyvalency, because new forms of labour segmentation can emerge : an elite of high skilled workers and technicians, de-skilled workers left over in the transition and a growing number of unemployeds (Hoss, 1986).

Secondly, managerial strategies designed to use labour more flexibly exist alongside structures and design assumptions which are still predominantly technicist and Fordist ; the way this contradiction is managed will exert a major influence on types of work organisation in the factories of the future.

4. New technology, the service sector and non-manual work

Here, I propose to focus primarily on the major occupational category of clerical work which has been the subject of a long debate in sociology. Consideration of managerial and professional employees has had to be excluded in this paper but clearly the use of information technologies to further organisational centralisation (or the reverse) has implications for the higher levels of the white-collar hierarchy (see, for example, Child, 1984). Likewise the position of clerical workers in manufacturing is not considered specifically here although similar trends could be identified in comparison with office automation in services.

Information technology applied to the office has developed over the last thirty years and forms part of a longer process of office mechanisation which began in the late 19th century. Traditionally offices have been less capital intensive than manufacturing but developments in microelectronics combined with telecommunications have enabled more rapid diffusion such that the paperless office is seen as part of a future scenario to complement the unmanned factory. However, the pattern of diffusion has been uneven between the various service sectors, with banking and insurance representing the highest pattern of use (Gershuny, 1983). This is reflected in research, where case study evidence is derived primarily from financial services, and to a lesser extent from retailing and medical services. The labour intensive nature of services combined with the rationalisation potential of office automation has (not surprisingly) led to an emphasis on the impact of technology on non-manual work which is likely to be more significant in the future than factory automation (Gill, 1985), with the qualification that the personal service element in many white-collar occupations will set limits to both the nature and extent of automation.

New technology applications in services are related to three central processes which are reflected in the office division of labour : collection, processing and retrieval/utilisation of information. These appplications include : automatic cash dispensers, customer information files, and electronic funds transfer for financial services ; in retailing, electronic point of sale systems are being developed ; in hospitals, medical records have been computerised, and in offices generally, word processors and electronic filing have been introduced. The current generation of computer systems are also being used to progressively integrate the different activities connected with the flow of information. On the one hand there is an increase in the use of mini and micro computers, and on the other, particularly in the larger organisations (banks, for example) there is an increase in 'networking' where 'real-time' computer systems allow access for input and retrieval of data from anywhere in the system. The earlier generation of 'batch computer systems' were, in contrast, analogous to NC machines in manufacturing—they were more likely to simply computerise previous manual activities and they were less 'interactive' based upon a category of workers whose task was purely one of data input via punch cards.

These developments have been interpreted as similar to the transition in manufacture from 'craft' to assembly and automation stages and consequently, broadly optimistic (scenarios) about changes in office work can be identified in the literature that bear a close resemblance to Blauner's conclusions. Thus it is suggested that interactive computer systems allow greater flexibility in work design the possibility of decentralisation of control and are-integration of work as the clerk and intermediate white-collar workers take on a wider range of tasks and are freed from the more routine office functions which are automated (Weir, 1977, Jones, 1980, Rajan, 1984, Wainwright and Francis, 1984).

However, directly contradictory evidence to the above in the form of extensive de-skilling of clerical work is also well documented. Bravarman's (1974) analysis of the office labour process depicts the same principles of scientific management applied to the office as in manufacture in which computerisation routinises work, separates conception from execution and leads ultimately to the proletarianisation of office workers who become part of a homogeneous mass of wage labourers. Subsequent empirical research has supported the labour process thesis of de-skilling with reference to the use of 'word processors' (Barker & Downing, 1980, Arnold SPRU, 1982) and office computerisation (Glenn & Feldberg, 1979). Similarly the ILO survey of research (Wernecke, 1983) into microelectronics and office jobs reported a predominance of international studies supporting the de-skilling argument. The recent British case studies by Crompton and Jones (1984) in banking, insurance and local government also found no support for the proposition that on-line computer systems reverse de-skilling ; over 90% of lower level non-manual jobs were found to be de-skilled or routine in content. At the same time, the authors did not concur with Bravarman's thesis of increasing 'homogenisation', instead, the office division of labour was polarised in terms of gender—women were the prime incumbents of lower level tasks whereas male clerks were more likely to be promoted.

As with manufacturing, therefore, the evidence about changes in the quality of non-manual work is contradictory—partly because of variations in the office labour process between sectors and nation states, and because clerical work has never denoted a homogeneous category upon which the consequences of computer use could be expected to follow a uniform pattern. However, contradictory research findings are also related to the ways in which 'skill' has been conceptualised and the wider theoretical perspectives about technology and work described in the introduction. General arguments from the 'human capital' perspective have frequently cited changes in the occupational structure, re-grading and formal qualifications as evidence of re-skilling ; whereas the de-skilling argument has focused upon objective changes in work content such as the amount of discretion and task variety, which, when reduced, imply job degradation. The evidence is further complicated by the fact that many clerical tasks were effectively 'de-skilled' prior to computerisation (Mills, 1956) as a result of Taylorism applied to the office and the 'Babbage' principle. The 'skill problem' will be discussed in the conclusion but, first, we can examine recent research (often from the 'contingency perspective') of a comparative nature which attempts to explain both de-skilling and trends towards polyvalency or multi-skilling.

Cross-national comparison of research findings reveals interesting variations in the way white-collar work is changing. Lane (1985) compares West German with British and American studies where, in the former, de-skilling was less in evidence. A variety of W. German studies point to the continuance of a minority of low-qualified workers who were de-skilled prior to computerisation and a larger group of qualified 'experts' ('sachbearbeiter') who have completed a formal apprenticeship. The former group is declining with office automation and the middle-level employees have experienced both up-grading and a re-integration of work tasks, in sectors such as banking and insurance. The apparent contrast between W. Germany and Britain is explained via a number of contingencies unique to the German case (Lane, 1985, Kocka, 1981) : the higher skill profile of clerks who have undertaken a formal apprenticeship, the stronger labour market position of qualified German clerks and managerial strategies aimed at flexibility in the face of market competition.

Similarly, the author's research into banking and hospitals in Britain and Switzerland (Charles, 1984, 1985) found evidence of polyvalency among Swiss qualified clerks. However in the case of Swiss banks, this occurred alongside a larger proportion of routine clerical workers largely women and some immigrant workers, who were performing only data input functions. This group had become a non-career grade of lower-level employees in an office division of labour which was becoming increasingly polarised. Contrastingly, in hospitals, the use of computer systems for medical records and patient admissions had not significantly changed the clerical division of labour and computers were supplementary to tasks which involved a high level of patient contact.

Finally, in a number of comparative studies in the service sector Child (1985; Child & Tarbuck, 1986) examines the variations in service jobs between, banks, hospitals and retailing with the introduction of new technology. De-skilling was found to be more prevalent in banking and retailing sectors than hospitals due to a number of factors : the ability of different work groups to influence the introduction of new technology (professionals in hospitals had greater autonomy), the characteristics of the tasks (routine tasks facilitate automation), institutional characteristics (particularly features of labour and product/elient markets) and organisational culture.

In sum, the changes taking place with regard to non-manual work are similar in many ways to those in manufacturing but the contradictions apparent in the philosophy of computer applications are possibly more transparent-strategies to cheapen labour and de-skill co-exist with traditional employer reliance upon the cooperation of non-manual employees, bureaucratic career structures and relatively higher trust relationships which mark the manual/non-manual divide and the position of non manual workers in the class structure.

5. Conclusions

Statements about general trends in the quality of work cannot be easily deduced from the current state of research : the post-industrial and human capital approaches have been largely discredited due to an uncritical acceptance of technology as a neutral force of production with its own logic and because the empirical evidence reviewed in the preceding discussion does not support it. The earlier versions of labour process theory which posited a logic of capitalist de-skilling have been seen to be over-simplified and subsequent research both from Marxist and contingency perspectives has moved towards explaining variations in the labour process and incorporating analysis of labour markets, managerial strategies, the role of organised labour into comparative research. The danger of the current spate of case studies is that the disparate range of empirical findings creates both difficulties of comparability of the research and a largely indeterminate view of the changes underway.

One of the major problems of comparability lies with the use of the term 'skill'. Firstly, definitions of skill can refer to objective or subjective characteristics of individuals, occupations or job requirements. 'Real' skills include reference to features of work such as unity of conception and execution, task range and complexity, amount of discretion, the amount of training required and the knowledge necessary to perform the task. For labour process theory these features have been placed in a historical context in which 19th century craft forms the model for skill. The subjective concept of skill emphasises the importance of employer's grading structures or trade union bargaining power in labelling certain jobs or occupations as skilled (Turner, 1962) irrespective of their 'real' skill content. The significance of occupational status needs, therefore, to be acknowledged in assessing changes in technical skill and the correspondence or otherwise between objective and subjective skills examined (Littler, 1982). Similarly, the concept of 'tacit skills' (Manwaring & Wood, 1985) which draws attention to workers' ingenuity and practical learning in all tasks, such that de-skilling is a matter of degree rather than an 'all or nothing' contrast with a romanticised view of earlier crafts. It is clearly important, therefore, to distinguish the different ways in which skills may be changing; notably, in the recent discussion of polyvalency, where it is necessary to specify whether there are new skills emerging, or a re-combination of several de-skilled tasks, or genuine multi-skilling.

Secondly, there is a need to differentiate between levels of analysis with reference to skill changes (Lee, 1982). The enterprise level studies reported here reveal important changes in the quality of work in both manufacturing and service sectors : new production concepts in the face of changed market conditions leading to more non-price competition in some industries are leading to an emphasis on flexible labour use, and multi-skilling for certain categories of work. At the same time, much routine work in manufacturing or services has either been de-skilled or automated out of existence leading to a polarised labour force on the basis of gender, skills and employment opportunities more generally. Current research is insufficient to estimate net effects on skill and employment levels. Qualitative changes in work at enterprise level need to be related to changes in employment at the industry level. Clearly, changes in technology cannot be taken in isolation as an 'independant variable', the outcome of managerial strategies and worker resistance is taking place in the context of recession, structural change in the economy and the international division of labour to structure employment opportunities. For example, it is likely that women in the service sector returning to employment after childbirth may suffer de-skilling, similarly, displaced traditional craft workers may not find alternative employment where their skills are utilised. The employment issue is outside the scope of this paper but the need to relate changes in the quality of work to the politics and economics of employment is developing in, for example, the use of long-wave theory mentioned previously.

Finally, the following topics are suggested as important in the research agenda for new technology and work :

1. There is a need to develop systematic comparative research both internationally and between sectors to determine under what conditions de-skilling and/or multi-skilling is likely to occur.
2. We know relatively little about workers' perceptions to technology at the workplace or more generally and how these vary by gender or occupation.
3. Attention has been directed more to the effects of new technology and less to the sociology of technology design and science in general.
4. Focus upon 'skill' is too narrow in terms of changes in work organisation. The analysis of the quality of work needs to include research into career patterns, intensity of work, the growth of part-time work, homeworking, shift work, employment opportunities and the technical division of labour.

REFERENCES

ARNOLD, E. (1982). *Microelectronics and women's employment in Britatn,* Sussex, Science Policy Research Unit.

BARKER, J., & DOWNING, H., (1980), "*Word Processing and the transformation of patriarchal relations of control in the office*". Capital and Class, 10.

BELL, D. (1974), *The Coming of Post-Industrial Society,* London, Heinemann.

BELL, M. (1972). *Changing Technology And Manpower Requirements in the Engineering Industry*, Sussex, University Press.

BLAUNER, R. (1964), *Alienation and Freedom*, Chicago, University Press.

BLUM, U. (1986) *Die Zukunft Der Arbeit*, Paper on IG Metal Survey to RKW Congress, Munich.

BRAVERMAN, H. (1974) *Labour and Monopoly Capital*, N. York, Monthly Review Press.

BRIEFS, U. (1980) *Arbeiten ohne Sinn und Perspektive* ? Cologne.

BRIGHT, J. (1985), *Automation and Management*, Boston, Harvard.

BURAWOY, M. (1979) *Manufacturing Consent*, Chicago. University Press.

CHARLES, A. (1984), "*New Technology and Clerical Work*". Paper to ISA Futures Research Committee, Zurich.

CHARLES, A. (1986), "*New Production Systems—Implications for work and training in the factory of the future*", Report of FAST research to the EEC, Turin.

CHILD. J. (1972), *Organisation Structure, Environment and Performance, the Role of Strategic, Choice*, Sociology, vol. 6, no.1.

CHILD, *et al.* (1984), "*Microelectronics and the Quality of Employment in Services*", in Marstrand, P. (ed.) (1984), *New Technology and the Future of Work and Skills*, London, Frances Pinter.

CHILD, J. (1985), "*Managerial Strategies, New Technology and the Labour Process*". in Knights, D., *et al*, (1985), *Job Redesign*, Aldershot, Gower.

CHILD, J., & TARBUCK, M., (1985), "*The introduction of new technologies: managerial initiative and union response in British banks*". Industrial Relations Journal, vol. 1, no. 3.

CHILD, J., (1986), "*Technology and Work*", *in Grootings, P., ed., (1986), Technology and Work, London, Croom Helm.*

COOLEY, M. (*1980*), *Architect or Bee* ?, Slough, Langley Technical Services.

KERN, H. & SCHUMANN, M. (1970), *Industriearbeit und Arbeiterbewusstein*. Frankfurt.

KERN, H., & SCHUMANN, M. (1984), *Das Ende der Arbeitsteilung* ? Munich.

KERR, C., *et al.*, (1970) *Industrialism and Industrial Man*, London, Heinemann.

KOCKA, J. ed. (1981) *Angestellte in Europaischen Vergleich*, Gottingen

LANE, C. (1985) "*White-collar workers in the labour process : the case of the Federal Republic of Germany*", Sociological Review.

LAWRENCE, P., & LORSCH, J. (1967) *Organisation and Environment*, N. York, Harvard.

LITTLER, C. (1982) *The Development of the Labour Process in Capitalist Societies*. London, Heinemann.

LEE, D. (1982) *Beyond De-Skilling-skill, craft and class*, in Wood S. ed. (1982) *The Degradation of Work?, London, Hutchinson*.

MANWARING, T., & WOOD, S. (*1985*) *The ghost in the labour process*, in Knights, D. op cit.

MARTIO, V. (1986) "*The role of the parties involved in the introduction of new technology*" Paper at the EEC Symposium, New Production Systems, Turin.

MILL, C. (1985) *White Collar*, N. York, Galaxy.

NOBEL, D. (1979) *America By Design*, Cambridge, M.I.T.

PAHL, R. (1984), *Divisions Of Labour*, Oxford, Basil Blackwell.

PENN, R., & SCATTERGOOD, H. (1985) "*De-skilling or enskilling ?: an empirical investigation of recent theories of the labour process*, British Journal of Sociology, vol. 36, no. 1.

RAJAN, A.(1984) *New Technology and Employment in Insurance, Banking and Building Societies*, Institute of Manpower Studies, Gower.

SABEL, C. (1982) *Work and Politics*, Cambridge, University Press.

SORGE, A., HARTMANN, M., WARNER, M., & NICHOLAS, I., (1983), *Microelectronics and Manpower in Manufacturing*, Aldershot, Gower.

THOMPSON, P. (1983) *The Nature of Work*, London.

TOURAINE, A., (1962) *An historical theory in the evolution of industrial Skills*, in Walker, C., ed. (1962) *Modern Technology and Civilization*, N. York, McGraw-Hill.

COOMBS, R., (1985), *Automation, Management Strategies and Labour Process Change*, in Knights, D., *et al*, (1985), *Job Redesign*, Aldershot, Gower.

CROMPTON, R., & JONES, G., (1984), *White-Collar Proletariat-De-Skilling and Gender in Clerical Work*, London, Macmillan.

EEC Commission, Directorate General for Employment, Social Affairs and Education, (1986), *Social Europe-New Technology and Social Change*, Luxembourg.

EDWARDS, R., (1979), *Contested Terrain : The Transformation of the Workplace in the 20th Century*, London, Heinemann.

FADEM, (1984), 'Automation and Work Design in the U.S. : Case Studies', in Warner, M., ed., (1984), *Microprocessors Manpower and Society*, Aldershot, Gower.

FRIEDMAN, A. (1977), *Industry and Labour*, London, Macmillan.

GALBRAITH, J. (1967), *The New Industrial State*, London, Penguin.

GALLIE, D. (1978) *In Search of the New Working Class*, Cambridge, University Press.

GALLIE, D. (1985), "*Social change and economic life : the current state of research*", Tbe Quarterly Journal of Economic Affairs, no. 1.

GERSHUNY, J. & MILES, I. (1983), *The New Service Economy*, London, Frances Pinter.

GERSHUNY, J. (1985), "*New Technology : What New Jobs* ;" *Industrial Relations Journal*, Vol. 16, no. 1.

GILL, C. (*1985*), *Work, Unemployment and the New Technology*, Cambridge, Polity Press.

GLENN, E. & FELDBERG, R., (1979), "*Proletarianizing Clerical wark : Technology and Organisational Control in the Office*, In Zimbalist, A. ed. (1979), *Case Studies on the Labour Process*, N. York, Monthly Review Press.

GOLDTHORPE, J. & LOCKWOOD, D., *et. al.*, (1968) *The Affluent Worker : Industrial Attitudes and Behaviour*, Cambridge, University Press.

GROOTINGS, P. (1986), *Technology and Work-East/West Comparison*, London, Croom Helm.

HOSS, D. (1986) *Technology and Work in the two Germanies*, in Grootings p., *op. cit.*

IG METALL, (1984), *Der Mensch muss bleiben*, Frankfurt.

JONES, T. (1980), *Microelectronics and Society*, Open University Press.

TURNER, H. (1962) *Trade Union Growth, Structure and Policy : a comparative study of the cotton unions*, London, Allen & Unwin.

WAINWRIGHT, J., FRANCIS, A. (1985) *Office Automation, Organisation and the Nature of Work*, Aldershot, Gower.

WEIR, M. (1977) *Are computer systems and humanised work compatible* ? in Ottoway, R. ed. (1977) *Humanising the Workplace.*

WERNEKE, D. (1983) *Microelectronics and Office Jobs*, Geneva, ILO.

WILKINSON, B. (1983) *The Shopfloor Politics of New Technology* London, Heinemann,

WILLIAMS, R. & STEWARD, F. "*New technologyagreements : an assessment*, "Industrial Relations Journal, Vol. 16, no. 3.

WOBBE, W. (1982) *Automobilarbeit and Roboterproduktion*, Berlin.

WOOD, S. ed. (1982) *The Degradation of Work*, London, Hutchinson.

WOODWARD, J. (1965) *Industrial Organisation*, London, Oxford University Press.

2

New Technology and Variations in Clerical Work

Tony Charles*

Introduction

Information technology applied to the office now represents a process of development spanning more than thirty years and part of a longer history of office mechanisation which dates back to the late 19th century. Compared to the historical and sociological analysis of the factory system, the history of office mechanisation and work processes is less adequately documented. The taxonomies of factory technology are well known, and include those of Blauner, Woodward, Touraine, Bright as well as Marx's pioneering work in *Capital*, Vol. I. Similar studies are rare in relation to the office, although comparisons are frequently made with factory production and processes analogous to the transition from machine tools, through assembly-line production to automation identified in the modern office. The different pessimistic and optimistic appraisals of the impact of automation on factory work are duplicated in studies of office automation and the subject of this paper is an examination of the debate about new technology and the nature of clerical work.

*Dept. of Sociology, Staffordshire Polytechnic, U.K.

The topic is approached, firstly, by a review of the classic studies of C.W. Mills and Lockwood in the 1950s about the 'new middle-class' compared with the more recent studies of office automation. This is followed by a critical analysis of the debate about clerical proletarianisation and deskilling which has informed most of the sociological discussion about office rationalisation and clerical work since the 1930s. Some of the problems engendered by this debate are identified such as the nature of 'clerical skills', the dangers of over-generalisation due to the variations in clerical work situations between industries and the position of clerks in the labour market, and the significance of the most recent developments in computer technology. The final section of the paper draws on the author's own research to suggest ways in which further studies can approach these issues, and argues for the need to demystify the new technologies by examining change in technology as neither 'neutral' nor inevitable in its consequences but closely related to the politics of the employer-employee relationship and the division of labour in the office as well as managerial ideology.

Lockwood and Mills "The New Middle-Class"

> "Mechanised and standardised work, the decline of any chance for the employee to see and understand the whole operation, the loss of any chance, save for a very few, for private contact with those in authority—these form the model of the future." (C.W. Mills, 1956)

> "The intensive mechanisation of clerical tasks, though comparable in its most advanced form to factory mechanisation has not played a particularly important role in the rationalisation of the work situation......" (D. Lockwood, 1958)

Many of the current arguments in Sociology related to office automation and clerical labour were anticipated in the 1950s and accounts of office rationalisation (which preceded mechanisation) appeared by the 1930s. A summary of the long and continuing debate about the class situation and 'class consciousness' of the 'middle-layers' is unnecessary in the space of this paper ; instead, I will concentrate on the ways in which technology has been seen to influence the clerical work situation. The work of Lockwood and Mills was probably the most influential of the early post-war studies of "the new middle-class". At that time the commercial application of computer technology was just beginning but office mechanisation had developed, particularly in the larger organisations such as banking and insurance, since the turn of the century.

Lockwood analysed the technology of the modern office in the context of what he called "the work situation" of clerical labourers and concluded that the process of rationalisation ushered in by bureaucratisation and mechanisation had so far only affected a minority of offices in England. Office rationalisation was most advanced in large corporations where greater isolation of clerks from management and the growing impersonality of the work situation was making the clerks work more like that of the factory operative. But work in the average office still bore more resemblance to the old 19th century counting house than the factory. Machinery was predominantly an aid to existing clerical functions (the factory demonstrated the reverse) although computerisation could, in the future, replace functions. Thus Lockwood concluded that despite a general decline in clerical status, and the gradual growth of the bureaucratic office, the manual/non-manual distinction was still of decisive importance for the class structure.

C.W. Mills' analysis of white-collar workers shared Lockwood's Weberian orientation to class, but although written seven years earlier, it emphasised already the routinised and de-skilled nature of office work in American society. Mechanisation was part of the new office division of labour which was steadily eroding the differences between office work and factory labour, and destroying the craft-like characteristics of work in the old office of the 19th century. These conclusions were different to those of Lockwood and reflected the earlier and more rapid development of both office mechanisation and a bureaucratised work situation in the USA compared to England ; in other respects their conclusions were similar—pointing to the persistence of differences in class consciousness and action between manual and non-manual workers. Together, the studies of Mills and Lockwood formed the dominant points of reference for the subsequent debate about the middle-class in general and clerks in particular. Lockwood's picture of the clerical work situation was not seriously modified until more than fifteen years later with the resurgence of Marxist interest in the class position of non-manual workers and proletarianisation.

Clerical proletarianisation and the 'de-skilling' debate

The main axis of debate today about clerical workers has not fundamentally altered since Lockwood's confrontation of the Marxian notion of 'false-consciousness' with an analysis of the class situation of clerks using Weberian categories. Conceptually, the language of 'work-situation' has been partially eclipsed by that of 'labour-process', and empirically, the developments in microelectronics have opened up a new chapter in the process of office mechanisation. The literature on the labour process is widely known and comprehensively reviewed elsewhere—most notably in the publication and subsequent debate about Braverman's 'Labour and Monopoly Capital' (1974). Accordingly, only Braverman's analysis of clerical labour will be discussed here, in the context of subsequent empirical research in Britain.

Whereas Lockwood concentrated on the transition from paternalistic to bureaucratic work situations and employment relationships, Braverman directs attention to the impact of Taylorism in the office and its affects on job-design. Braverman asserted what Mills thought situation encompass both de-skilling and a loss of elements of control as computerisation has resulted in a centralisation of information away from the clerks themselves. In terms of class analysis, to the extent that clerks previously exercised both 'capital' and 'labour' functions the office rationalisation has tended to affect both aspects of the clerk's role. These changes are further experienced by clerical workers as a sense of subordination to the demands of the computer (particularly with batch systems) and help to explain the increase in trade union density in the GB finance sector during the 1960s and 1970s. However, in criticism of Braverman, it is argued that these changes have resulted in further polarisation in white-collar work (particularly along the lines of gender inequality) rather than tendencies towards homogeneisation of the labour force. The counter arguments appear in three forms, and these will be briefly discussed before proceeding to a number of critical comments about the state of current research.

Firstly, using very different evidence, Stewart argues that high rates of social mobility found in a large survey of clerks suggest that no clerks have been proletarianised. Distinguishing persons from structural positions research (Stewart, 1980) reveals that the occupational category of 'clerk' is highly heterogeneous containing high rates of inward and outward mobility—male clerks are still likely to experience promotion into management or will move to other employment. Secondly, Giddens (1981), in his assessment of the 'new middle-class' debate, finds little support for the thesis of proletarianisation. Evidence is cited which points to the continued significance of the manual/non-manual divide and to the absence of de-skilling. Computerisation has tended to reduce the amount of routine work and require more qualified personnel. On balance, automation in the office results in more up-graded posts than down-graded ones, and, in common with Lockwood's earlier argument, Giddens argues that office mechanisation as a whole has not led to factory type 'machine-minding'—office machines tend, on the contrary, to replace routine manual functions. This argument is pursued in Giddens (1982) later critique of Braverman in which it is argued that labour process theory is both over-deterministic and over-simplifies the process of technological change. Thirdly, Mumford, writing from a 'socio-technical systems' perspective, questions the inevitability of Taylorism applied to systems design. Case study evidence is used to demonstrate that user participation in the design process can reconcile both demands for technical efficiency and the need to humanise work-computer systems are sufficiently flexible to allow a variety of design philosophies (Humford, 1980).

What conclusions can be drawn from this brief comparison of the arguments ? In the first place, the case study evidence is contradictory in relation to the de-skilling/upgrading issue. Secondly, labour process theory has concentrated on objective changes in the nature of work whereas the counter arguments have focused upon different evidence to do with labour markets, social mobility and the reactions and perceptions of clerks towards technical change. The result has been to identify a number of important weaknesses in the Braverman's thesis of clerical de-skilling, elaborated both by his critics and others generally supportive of labour process theory. The explanatory power of this theory lies in its 're-discovery' of the significance of capitalist control in the production processes and work design ; technology and contol are not neutral processes but class based features of the division of labour. This perspective was highly influential in the re-appraisal of 'conventional' industrial sociology but was, itself, not without criticisms—both theoretical and empirical. Despite the historical focus on the labour process, critics have revealed the deterministic implications of the 'functionalist' Marxism derived from class analysis which has informed the study of the labour process : the imperatives of capitalist control in the labour process appear uncontested and the inevitable product of managerial design ; an approach which leaves little 'theoretical space' for workers' reactions to technological change, in which strategies are devised to counteract the impact of Taylorism on the shop-floor. Likewise the concentration on the role of scientific management overlooks other factors of significance : alternative forms of control (bureaucratic, 'responsible autonomy') and the role of labour markets and the state as they impinge on the workplace (Edwards, 1979).

Turning to clerical work in particular, there is ambiguity both about the nature of 'clerical skills' and the historical and comparative basis for de-skilling.

A note on clerical skills

The model for changing skill levels has usually been based upon the industrial worker with the traditional craftsman as the historical benchmark. 19th century craftsmanship has also provided the model for clerical work both for Mills and Braverman. Lockwood's description of the 19th century office refers to the need for a general education and literacy, combined with good arithmetic ability and accurate work. However, the abilities associated with the various office functions were probably less important than what may be called 'social skills' and a special relationship to the employer-abilities to deal with people (clients) and symbols, rather than meterial things : qualities seen by employers as responsibility, trust, diligence, confidentiality and so on. These attributes were, in turn, seen as promotion qualities and thus separated higher grade clerks from the lower grades who performed more routine functions. Historical evidence thus reveals variations in the types of clerical work—much was of a routine nature but aspects of the clerk's role which involved greater discretion would probably be now labelled 'cadre' or managerial in function. Variations both in terms of task variety and discretion are also apparent today within the broad category of clerical employment. Tendencies to de-skilling are less amongst those in the office hierarchy who retain a significant amount of client contact as part of their work role, are in a position of higher 'trust' with the employer and/or possess specialised knowledge (particularly when strong professional associations defend occupational interests).

The implications of this argument for the de-skilling debate are twofold : firstly, the tendency to romanticize 19th century clerical work according to the craft ideal. In the same way that industrial craftsmen were never a majority in the 19th century, clerks also comprised a heterogeneous category in which many performed routine manual functions. Similar ambiguities exist in the comparison of more recent changes in clerical work involving computerisation. Prior to office mechanisation the processes of rationalisation and fragmentation of work had already begun, but were limited in their effects by the size of the typical office. However, prior to the widespread use of information technology, many 'manual' office tasks were already de-skilled by the use of less sophisticated office machines. Secondly, the definition of clerical skills and even of clerical work itseif has been seen to be problematic. The definition of clerical skills encounters the same difficulties as attempts to define 'non-manual' work in terms of its intrinsic characteristics—many clerical tasks involve manual operations and conversely, all manual labour involves some mental effort. In order to overcome these kind of problems a wide range of criteria (Bain, 1972) (including - work milieu, style of dress, proximity to authority, and function of the work) have been identified as a basis for differentiation. What, then are 'non-manual' skills ? Skills in general refer to two characteristics—one intrinsic to the job, denoting the training involved, and the level of task variety and discretion required ; the other concerns the 'negotiated character' of skills in which claims to skill depend upon the bargaining power of different occupational groups and their position in the labour market. Features such as job knowledge and discretion may therefore be necessary but not sufficient conditions for claims to skill—other factors such as the labour market and workplace bargaining also play a significant part in the formation of skill as a social category. Furthermore, the importance of client relationship and the general significance of wider 'social skills' referred to previously in relation to some types of clerical work, call in to question the applicability of the term 'skill' itself since its meaning, in English usage, relates primarily to industrial work. French and German research on 'de-skilling' has used the broader term—'dequalification' for 'entqualifizieren' which denote changes in 'qualifications' for a particular job in the broadest sense of the term-that is to say, qualities involved in work (including paper qualifications). The German 'geschicklickeit' or the French 'habilite', as LITERAL translations of 'skill', are inappropriate since they refer to manual abilities such as dexterity but not to occupations.

Returning to the general issues raised by the 'de-skilling' debate—a review of the empirical evidence about computerisation and clerical work in both Europe and the USA clearly reveals that the weight of evidence is in favour of de-skilling rather than upgrading (Forester, 1980, Spru, 1982). Whether, skill is defined narrowly in terms of technical features of office tasks or whether we talk of dequalification, the large majority of research findings indicate increased rationalisation and fragmentation of office tasks, centralisation of managerial control and generally declining conditions of work and promotion prospects for routine clerical employees. The problems outlined above concerning definitions of skill and historical comparison do not refute the thesis of de-skilling, they indicate errors or omission and the need for more cross-cultural and inter-industry comparisons in order to assess the significance of variations in clerical work situations. We shall return to this point in the concluding section which discusses recent studies, including comparative case studies from the author's research.

However, the counter arguments to de-skilling in the form of the Cambridge research reviewed above are not sufficient as a basis for critique of labour process theory. The evidence about social mobility of male clerks, whilst indicating the continuance of upward routes into management for a minority of qualified males, did not investigate the situation of the clerical majority—female employees. Gender based inequalities at work and in the labour market serve to stratify the clerical workforce into minority of male cadre and specialist positions and a majority of female clerks who perform more routine work and experience inferior career chances. Social mobility research in Britain has not adequately examined gender stratification and its relationship to class analysis as far as clerical workers are concerned (a point which applies equally to the Braverman's thesis). Of course, the social mobility studies were not concerned with de-skilling as such, but the evidence about careers of male clerks does not support the argument of *general* upgrading of clerical employees in recent years. Similarly, Gidden's (1981) argument for clerical upgrading was based upon a very limited selection of case study evidence, (now quite dated) primarily from the USA. In addition, the evidence about job upgrading is itself inconclusive since tasks may well be re-graded or re-labelled without any significant change in the type or level of work undertaken. Nevertheless the Cambridge research as a whole has made a more important contribution to the wider debate about proletarianisation by calling into question the tendency in Marxist analysis to 'read off' changes in class consciousness and action as more or less automatic outcomes of objective class position and the labour process associated with employment in the 'middle-layers' of the class hierarchy. A detailed discussion of the proletarianisation thesis is outside the scope of this paper ; in any case, de-skilling or 'dequalification' is neither a necessary nor sufficient condition for proletarian status (as Crompton notes, skilled workers can supposedly be both skilled and Proletarian). To sum up this review of the arguments for and against de-skilling, recognition of the limitations of labour process theory does not entail acceptance of the counter arguments about clerical upgrading consequent upon technological change.

INFORMATION TECHNOLOGY AND CLERICAL WORK : CASE STUDIES FROM BANKING AND HOSPITALS

The expansion of white-collar work in general and clerical employment in particular constitutes a major change in the UK occupational structure and is typical of similar growth patterns in other advanced capitalist societies. White-collar workers accounted for 42% of the occupied population in Britain in 1971 (compared to 18% in 1911) and clerks within this category totalled 14% of the occupied population in 1971 (4.5% in 1911) (Price, 1976). This growth in clerical numbers was nearly all accounted for by an increase in female employment. The term 'white-collar' has, however, never denoted a meaningful social category in class terms since it includes professions as diverse as doctors and managers at one extreme and routine clerical grades and service personnel at the other. Even the clerical occupational category conceals important variations both in the nature of work and conditions of employment.

Our research aimed to examine these variations, using case studies from different sectors of clerical employment and by the comparison of the same sectors in different societies (Charles, 1983). Banking and hospitals were chosen as two markedly different sectors for clerical work, in terms of organisational culture, nature of work and position of these institutions in the wider economic and political structure. Banking and insurance are already more highly automated than the rest of the tertiary sector and the larger institutions represent highly bureaucratic and centralised structures. The Swiss banks in the study were also more advanced in the use of information systems than their British equivalents, having moved earlier to 'real-time' integrated computer systems from the earlier 'batch' computer applications. The British and Swiss hospitals in the survey were broadly similar in terms of the degree of automation of clerical and administrative tasks. Patient registration and admission functions represent the most developed applications for computerisation and they largely affect the clerical staff. Other systems for laboratory work and medical diagnosis are less widely in use at the moment. Compared to banks. hospitals, although subject to centralised control in terms of finance (in Britain through the regional structure of the National Health Service, in Switzerland, through the Cantonal authorities), are more decentralised in their use of information technology reflecting the relative autonomy of departments and of the medical profession.

Banking is a clear example of the changes Lockwood described from paternalistic to bureaucratic work situations ; also of the process of dequalification. The old image of the banker as 'l' homme de confiances' has given way to the specialist and the routine clerk without banking qualifications. In the Swiss banks, for example, most clerks had completed a banking apprenticeship before 1950, this declined to 50% for all Swiss banks by 1970 ; in our survey of a large bank the percentage with banking qualifications had dropped to below 40% by 1984. The enormous increase in banking business since World War II led to a rapid increase in the employment of routine clerical workers and of female employees (an increase of 400% between 1950 and 1980). (Steffan, 1970. Hopflinger, 1980).

As a result, clerical employment in banks has remained relatively secure—increases in productivity have not been accompanied by redundancies. However, further automation is likely to reduce the need for routine clerical work, because the job of data-entry is subject to further rationalisation and the 'self-service' principle is extended (cash dispensers and so on). Thus, by 1984, employment levels in the large Swiss banks have remained constant and are likely to fall in the future. As routine grades are more vulnerable to these changes the result will most probably be a gradual 'de-feminisation' of the clerical workforce.

The processes of rationalisation and office mechanisation in banking began after the turn of the century and the first computer systems were introduced in the 1950s. Few clerical workers can, therefore, remember a 'manual' work situation without the computer, although many have experienced the transition from earlier batch systems to 'on-line' real time systems. The consequences for the nature of clerical work follow a pattern characteristic of other reports on the finance sector in Europe (Leclerq, 1983). Preliminary analysis of our data shows that virtually all clerks interact with the computer in their work. Some of the data-entry work (previously done by punch-card operators) has been decentralised—for example' to an on-line terminal by the cashiers in the branches, together with a large increase in the number of terminals in the bank's specialised departments. However, the routine work of data-entry and checking of accounts remained for the large volme of daily transactions. Work previously done on punch-cards was now entered on the terminals but this continued to be a fragmented and highly controlled set of routine operations. Work in this category (particularly in the head office computer centres) most resembled the 'factory office' and was organised along classic Tayloristic principles.

The nature of interaction between the clerk and the computer reflects the hierarchical organisation in the bank, as follows :

(*a*) The 'cadres'—have broad access to information on the computer for management functions and customer services, are in a position of 'high-trust' and use the computer as an information source.

(*b*) Specialists—divided into technical specialists who are separated in the organisation from the rest of the bank's staff and banking specialists who have restricted access to computerised information for particular customer services.

(*c*) Routine clerks—in a position of 'lower trust' with management, and in which a major aspect of the work is concerned with input of data connected with banking transactions.

A large proportion of staff in this third category do not have bank qualifications and their promotion opportunities are limited. However, this overall picture of the division of labour in banking conceals significant variations in the clerical work milieu. Office size is still one of the most important factors affecting the work situation. Smaller branches in the bank still retained elements of a paternalistic organisation, and clerical tasks were likely to involve greater job variety and more customer contact. The bank head office, on the other hand, contained a high proportion of staff engaged exclusively in terminal work which was subject to frequent evaluation of staff output and number of errors. The work was also more likely to be perceived by clerks as lacking in variety or discretion, causing stress and subject to close supervision. Attachment to work for routine female clerks was frequently described in terms of job security and conditions (compared to other employment possibilities such as shop work) rather than intrinsic aspects of the job. Work of this type in banking most clearly reflects the changed image of the banker—to quote a recent French research report—'Le jean remplace le complet veston'. (Leclerq, 1983).

The hospital case studies revealed a number of contrasts in terms of the impact of computerisation on clerical tasks. Computer applications developed later in hospitals compared to banks and their effect on clerical tasks was found to be minimal. Computerisation of medical records and patient admissions did not alter the clerk's contact with the patient, instead the hospital information system provided patient details and did not involve data-entry to the same extent as clerical operations in the banks. The level of clerical qualifications was no higher in the hospitals than the banks and promotion opportunities for lower grade clerks were as restricted as the banks, but the work milieu was different. Hospital organisational culture was permeated by the ethos of 'patient care' which filtered down from the medical profession to the clerks ; in this context, the computer system was seen more as an aid to clerical tasks and managerial control was based more upon 'responsible autonomy' than 'factory office' principles. Nevertheless, at the wider level, hospitals (whether private-funded or part of state welfare provision) are also guided by criteria of efficiency and productivity but their effect on work situation is mediated by professional control and relative autonomy of departments at the level of the hospital. (Gough, 1979. Jamous and Peloille, 1970).

The implications of this brief resume of case study evidence for the debate about new technology and clerical work are twofold : firstly, the tendency towards dequalification in clerical work does not follow a uniform pattern—variations are apparent in different industrial or commercial organisations. An adequate classification of clerical work contexts by sector is an important priority for future research in order to qualify the global optimistic or pessimistic scenarios about technology and the future of work. Secondly, the outcome of tendencies towards dequalification and office rationalisation is partly dependant upon the type of clerical work affected. The greater the amount of 'indeterminacy' and customer or client contact involved in clerical tasks the greater the possibility for clerks to retain areas of discretion in the work role and to resist a worsening of the job content (Jamous, 1970). Thus, in banks, cashiers, credit clerks and others in 'front office' jobs had tasks involving relatively more discretion and variety than clerks involved in 'back-office' functions such as the processing of accounts. Similarly; in hospitals, differences existed between the work of clinic receptionists and that of filing clerks.

Apart from variations in the work situation, national differences in terms of the role of the state, labour markets and the role of white-collar trade unions or staff associations also exert a considerable influence on the effect of technology in the office. Thus, in several European societies, clerical employees in the state sector have been more highly unionised : in Britain tendencies towards types of action and bargaining similar to manual unions exist and major white-collar unions are affiliated to the Trade Unions Congress. In Switzerland, white-collar unions (often with low membership densities) exist alongside a national professional association of white-collar employees which administers examinations for professional status and represents qualified clerks as a political pressure group. The history of white-collar organisations and their relationship to the labour movement in France and W. Germany again reveals important contrasts in terms of the politics of the labour market and the work situation(Kocka, 1981). The consequences of these variations for middle-class politics are well documented elsewhere; for the purposes of this discussion, it is sufficient to note that, as in manufacturing, clerical workers vary as to the power of their collective organisations in resisting dequalification and negotiating the introduction of new technology.

The way in which negotiations over the introduction of new technology have been institutionalised therefore reflects national and sectoral traditions of collective bargaining; but they are also influenced by the existing skill profile of the labour force and conditions prevalent in labour and product markets. Thus, in the Swiss banking case study the introduction of new technology was implemented from the top down in a highly centralised structure with negligible union participation. Negotiation was confined to the later stages of implementation and was confined to conditions of work such as health and safety. In addition, the bank's earlier practice of recruiting large numbers of unqualified staff for routine data-entry functions had created a polarised prior office division of labour, particularly in the regional data-processing centres. The predominantly male specialist workers (holders of a banking apprenticeship) frequently experienced an enlargement of their skills based upon more recent computer applications, whereas lower level ('mitarbeiter') clerical workers continued to perform de-skilled tasks which were job specific—presenting restricted opportunities for upward career movement or alternative employment.

Similar evidence from the banking sector in the UK is well documented in the recent studies by Child and Tarbuck (1985) and Crompton and Jones (1984). In the former, case studies revealed the weak bargaining position of routine clerks who constituted a "secondary labour force" and whose tasks facilitated automation and tightly rule-governed work procedures as the banks pursued a policy of rationalisation (in response to increased market competition and declining profitability on retail banking transactions). Based upon case studies of banking, local government and insurance, Crompton concluded that 90% of clerical jobs in her sample required no skill in terms of discretion and tasks were entirely rule-governed in a gender based dual career structure. It appears, therefore, that sector characteristics, as indicated in banking, override national differences as expressed in occupational training, education and collective bargaining. This is not surprising, given the international nature of capitalist markets in general and broadly similar state control over private banking operations in particular.

However, contrasting evidence from West German case studies of clerical work presents a different picture of current changes in clerical work organisation. In a comparison of British and West German research, Lane (1985) notes a number of contingent factors which are specific to the position of white-collar workers in West Germany : the existence of a strong skill tradition based upon formal vocational training formalised institutional arrangements for a measure of co-determination, and a relatively lower level of feminisation of white-collar employment. Citing several West German studies, Lane argues that the consequences of new technology for non-manual workers cannot be described in terms of a uniform logic of de-skilling. Routine clerical workers constituted a smaller fraction of white-collar employees and their work had been effectively de-skilled prior to computerisation; qualified specialists on the other hand had, in some cases, experienced a raising of skill requirements and job enlargement, following more recent applications of EDP where previously fragmented tasks are re-integrated.

The West German example illustrates the importance of comparative research for an analysis of the conditions under which the use of information technology leads to de-skilling, polyvalency (re-integration of tasks, which may or may not entail genuine re-skilling), or a combination of the two for different work groups. Two sets of factors appear to be important in determining the eventual outcome :

1. Managerial strategies for work organisation and control in the context of the nature of both labour markets and the markets for goods or final services.
2. The prior division of labour in the organisation and the bargaining position of employees.

For example, in the West German studies, the higher skill profile employees in banking and insurance, together with the need for flexibility in the face of market competition, resulted in more polyvalent jobs for specialists and increased automation of routine tasks. However, cross-national differences should not be exaggerated. Evidence from the European financial sector (Euro-Fiet, 1985) depicts a general trend towards polarisation of non-manual employees and poor employment prospects for routine clerks (who constitute a larger proportion of white-coller employees in Britain compared to West Germany). In Switzerland, despite a similar emphasis to West Germany upon formal banking apprenticeships, my research revealed, as noted previously, that the proportion of bank employees in the case study with an apprenticeship had declined and routine clerks consituted a majority of non-managerial employees, particularly in the larger branches.

There is also a third possibility which would also partly explain these national differences in the skill profile of non-workers : namely, the level of development of computer systems in banking in different countries. West Germany has gone further down the road of fully integrated information systems (although the use of automatic cash dispensers is higher in Britain) and therefore, it could be argued, British banks will follow a similar development with a consequent reduction in the proportion of routine clerks as they move to fully integrated real-time systems (Euro-Fiet, 1985; Rajan, 1984). Notwithstanding the inherent flexibility of current information systems, however, the routine job of data-entry remains and the resultant work design appears to be related more to managerial strategies and contingencies such as the nature of labour markets than a logical consequence of the technical capabilities of the computer system. The comparison of Switzerland and West Germany tends to support this conclusion. although further research would be necessary to provide reliable indicators about the technical features of the banking computer systems in the respective countries.

Conclusion

The preceding discussion of clerical work has been necessarily selective and restricted to branches of the service sector where information technology has been widely used. This reflects a pattern of diffusion of new technology which is uneven between sectors (Gershuny & Miles, 1983). Likewise, discussion has been limited to that part of the office division of labour directly concerned with service provision rather than the role of typists and secretaries : a similar debate about the implications of word processing for clerical workers is already well documented in the literature (spru, 1982 : Barker & Downing, 1980).

The de-skilling debate in relation to clerical work and the class structure is now informed by an increasing number of empirical studies which are beginning to rival the dominant concern with manual work and the factory in the sociology of work. The earlier versions of labour process theory which posited a general logic of capitalist de-skilling and proletarianisation were seen to be over-simplified and subsequent research both from Marxist and 'contingency' perspectives (Knights, 1985) has moved towards the explanation of variations in the labour process with reference to labour and product markets, managerial strategies, the role of organised labour and the significance of gender inequality in the office division of labour. The danger of the current proliferation of case studies is that the desperate range of empirical findings creates both difficulties of comparability of the research and a largely indeterminate view of the direction of change in clerical work. Nevertheless, the case studies from banking, hospitals and service sector employment reported here provide a clue to the major trends of change in clerical work and the likely pattern of variation in the office division of labour; these can now be summarised as follows :

1. The labour intensive nature of services combined with the rationalisation potential of office automation means that the impact of new technology on clerical employment and work content is likely to be more far reaching in the future than factory automation (Gill, 1985). A process of de-skilling (in terms of task complexity required in certain categories of clerical discernible, particularly with the earlier 'batch' computer systems. In the case of banking this had carried over to more recent 'real-time' computer system for the lower grades of clerical work. This process reflected the pre-existing office division of labour in large banks based upon Taylorist principles and a centralised, technicist system design philosophy. Many tasks had been effectively de-skilled well before computerisation.
2. At the same time, examples of new skills associated with programming and computer software development have created a new stratum of 'information workers' in the office hierarchy. There were also examples of 'polyvalency' or multi-skilling affecting particularly qualified 'experts' who interpret or use information rather than input data. In banking, an example of polyvalency from the Swiss case was the addition of customer advice to the traditional role of the cashier (which had changed due to the use of counter terminals and cash machines).
3. There are two possible explanations for the co-existence of both de-skilling and more flexible forms of the office division of labour with a more polyvalent task structure—one refers to the level of technological development ; the other places emphasis on the nature of market competition and the need for a co-operative and flexible workforce. The 'technological' explanation refers to the current generation of computer systems which progressively integrate the different activities connected with the flow of information. There is an increase in 'networking' where 'real-time' computer systems allow access for input and retrieval

> of data from anywhere in the system. The earlier generation of 'batch' systems were, in contrast, analogous to NC (numerically) machines in manufacturing—they were more likely to simply computerise previous manual activities, whereas CNC (computer numerical control) in manufacturing and current generation information systems in services are more interactive. They allow greater flexibility in work design, the possibility of de-centralisation of control and a reintegration of work as the clerk takes on a wider range of tasks and is freed from the more routine functions which are automated.

(Jones, 1980 ; Weir, 1977 ; Rajan, 1984 ; Wainwright & Francis, 1985). However, whether information systems are designed in this more flexible way is a matter of choice based upon managerial strategies (Child, 1985) and the ability of work groups to negotiate over job design, rather than a consequence of a supposed 'logic' of the technology as an independent variable.

The second explanation for the trend towards flexible organisation and multi-skilling relates to the contradictions inherent in managerial control and organisational design, particularly for the traditionally loyal non-manual employees. In the case of banking, the importance of reacting quickly to customer demand for services and increased competition between financial institutions place a premium on flexible organisation in addition to cost rationalisation. The reliance of management upon the 'tacit skills' (Manwaring & Wood, 1985) of clerical workers, the need to harness the skills of qualified specialists and the relatively higher 'trust' relationships characteristic of non-manual work serve to place limits upon strategies of de-skilling and continue to separate non-manual from manual workers.

The application of new technology in the hospital case studies provided a contrast with the financial sector. Computerised patient admission systems and medical records had little effect on the prevailing office division of labour or the quality of work. This was explained by the organisational culture in hospitals and the importance of a 'personal service' element for many clerical functions. Where clerical tasks continue to incorporate a significant amount of client contact, the 'social skills' that this requires will therefore set limits to strategies of job degradation.

Developments in white-collar work organisation towards flexible work groups, project and matrix forms of organisation and the 'after Japan' managerial emphasis on work culture and employee commitment are very similar to what Kern and Schumann (1984) have called 'a new production concept' in manufacturing. Recent evidence from metal machining points to a re-integration of tasks using FMS (flexible manufacturing systems) rather than Tayloristic forms of job fragmentation and a hierarchical division of labour in response to changed market conditions. The result is a search for greater flexibility in production compared with the 'Fordist' strategy of cutting production costs for standardised goods (EEC, 1986 ; Sabel, 1982). However the consequences of these changes for employment and the overall industrial skill profile of manual workers is a matter of dispute (Hoss, 1986)—one likely scenario is the emergence of a small group of multi-skilled workers and technicians, a remaining category of un-skilled workers whose jobs cannot be automated and a larger category of unemployed : similar to the possible employment consequences for clerical workers discussed below.

4. No examples of direct redundancy due to new technology were found in the case studies. Banking and hospitals were regarded by respondants as 'secure employment' with good conditions of service. In banking this reflected the traditionally paternalistic employment culture and an 'internal labour market' with a recognised career structure. The process of office rationalisation was managed by early retirement and natural wastage. However, the post-war period of expansion in clerical work has clearly ended and the future outlook in banking and public sector clerical work is one of labour contraction and declining career opportunities (Euro-Fiet, 1985).

Based on the case studies, the future employment scenario for clerical workers will reflect the process of polarisation in the office division of labour. Clerical work has never denoted a homogeneous category and the division between qualified career staff and non-career, routine clerks is likely to increase. With the further integration of activities through the current generation of computers and the gradual automation of routine office work, there will be a decline in employment of routine clerical workers which will primarily effect women workers. There is already a trend towards increased part-time employment in banking, for exemple.

It is difficult to draw the boundary between 'specialist' or qualified clerks and management since this group merge into 'cadre' functions in the bureaucratic career structure. As in manufacturing, therefore, the process of office rationalisation and automation creates 'winners' and 'losers' in the labour market : the latter constitute a growing 'secondary labour force' involved in part-time employment or unemployed. This office polarisation is structured around prevailing gender inequalities in the labour market.

5. Finally, it was noted that the way information technology is designed is inseparable from the prior office division of labour and managerial objectives. As in the history of factory technology (Noble, 1979), office technology has been implemented with minimal employee participation—particularly in the early design stage (Martino, 1986 ; Williams & Stewart, 1986). Systems design has, therefore, been predominantly 'technicist' rather than human oriented (Cooley, 1980). Nevertheless, a process of de-skilling was not the only one in evidence; but where there was a policy of using new technology to build upon clerical skills and use labour more flexibly this was restricted by design assumptions and previous forms of work organisation that are still predominantly Taylorist and bureaucratically centralised. The way this contradiction is managed and negotiated by employees will exert a major influence on the office division of labour in the future.

REFERENCES

ABERCROMBIE, N. & URRY (1938), J. *Capital Labour and the Middle-Classes*, London : Allen & Unwin.

ARNOLD, E. (1982), *Microelectronics and woman's employment in Britain*, Sussex, Science Policy Research Unit.

BAIN, G. & PRICE, R. (1972), '*Who is a white-collar employee ?*' Brit. Journal of Industrial Relations, vol. 10, no. 3.

BAIN, G. & PRICE, R. (1976), '*Union growth re-visited*' Brit. Journal of Industrial Relations, Nov.

BARKER, J., & DOWNING, H. (1980), "*Word Processing and the transformation of Patriarchal relations of control in the office*", Capital and Class, 10.

BRAVERMAN, H. (1974), *Labour and Monopoly Capital*, N. York, Monthly Review Press, 1974.

BURAWOY, M. (1979), *Manufacturing Consent*, Chicago, University Press.

CHARLES, A. & STANLEY, G. (1983), *The social implications of computerisation in a hospital out-patients department*, Report to the West Midlands Regional Health Service.

CHILD, J. *et al.* (1984), "*Microelectronics and the Quality of Employment in Services*", in Marstrand, P. (ed)., *New Technology and the Future of Work and Skills*, London, Frances Pinter.

CHILD, J. (1985) '*Managerial Strategies, New Technology and the Labour Process*", in Knights, D., *et al.*, *Job Redesign*, Aldershot, Gower.

CHILD, J.&TARBUCK, M. (1985), '*The introduction of new technologies: managerial initiative and union response in British banks*". Industrial Relations Journal, vol. I, no. 3.

COOLEY, M. (1980), *Architect or Bee ?*, Slough, Langley Technical Services.

CROMPTON, R. (1979), *Trade unionism and the insurance clerk*, Sociology, vol. 13.

CROMPTON R. & JONES, G. (1979), '*Clerical proletarianisation : myth or reality ?*, in Day, G. Diversity and Decomposition in the Labour Market, Aldershot, Gower.

CROMPTON, R., & JONES, G. (1984), *White-Collar Proletariat-De-Skilling and Gender in Clerical Work*, London, Macmillan.

E.E.C. Commission, Directorate General For Employment, Social Affairs and Education, *Social Europe—New Technology and Social Change*, Luxemburg.

EDWARDS, R. (1979), *Contested Terrain : The Transformation of the Workplace in the 20th Century*, London, Heinemann.

EURO-FIET (1985), (International Federation of Commercial, Clerical, Professional and Technical Employees) *Bank Workers and New Technology*, Geneva.

GERSHUNY, J. & Miles, I. (1983), *The New Service Economy*, London, Frances Pinter.

GIDDENS, A. (1981), *The Class Sructure of the Advanced Societies, London, Hutchinson.*

GIDDENS, A. (*1982*), *Profiles and Critiques in Social Theory*, London : Macmillan.

GILL, C. (1985), *Work, Unemployment and the New Technology*, Cambri- Polity Press.

GOLDTHORPE, J. (1980), *Social Mobility and Class Structure in modern Britain*, Oxford : Clarendon Press.

GOUGH, I. (1980), *The Political Economy of the Welfare State*, London : Macmillan.

HOPFLINGER (1980), *Die Anderen Gewerkschaften* Zurich : Eco-Verlag.

HOSS, D. (1986), *Technology and Work in the two Germanies*, in Grootings. P. (ed)., *Technology and Work : East/West Comparison*, London, Croom Helm.

HYMAN, R. & PRICE, R. (1983), *The New Working Class* ? London : Macmillan.

JAMOUS, H. & PELOILLE, B, (1970), *Change in the French University Hospital system*, in Jackson, J. (ed), *Professions and Professionalisation*, Cambridge.

JONES, T. (1980), *Microelectronics and Society*, Open University Press.

KERN, H., (1984), & Schumann, M. *Das Ende der Arbeitsteilung*? Munich

KLINGENDER, F. (1935), *The Condition of Clerical Labour in Britain*, London : Martin Lawrence.

KNIGHTS, D. et al (eds) (1985), *Job Re-Design*, Aldershot : Gower.

KOCKA, J. ed. (1981), *Angestellte in Europaischen Vergleich*, Gottingen.

LANE, C. (1985), "*White-collar workers in the labour process : the case of the Federal Republic of Germany*", Sociological Review.

LECLERG, R. (1983), *Les Conflits du secteur bancaire*, Sociologie du Travail, no. 1.

LOCKWOOD, D. (1958), *The Blackcoated worker*, London : Allen & Unwin.

MANWARING, T. (1985), & Wood, S. *The ghost in the labour process*, in Knights, D. op. cit.

MARTINO, A. (1986), "*The role of the parties involved in the introduction of new technology*" paper at the EEC Symposium, New Production Systems, Turin. FWST, Brussels, Belgium.

MILLS, C. (1956) *White Collar*, N. York, Galaxy.

MUMFORD, E. & BJORN—ANDERSON, N. (1980), *The Human Side of Information Processing*, The Hague : North Holland.

NOBLE, D. (1979), *America by Design*. Cambridge, M.I.T.

RAJAN, A. (1984), *New Technology and Employment in Insurance Banking and Building Societies*, Institute of Manpower Studies, Gower,

SABEL, C. (1982), *Work and Politics*, Cambridge University Press.

STEFFAN, H. (1970), *Das Bankpersonal Heute und Morgen*: Bankwirtschaftliche forschung, Band 3, Berne.

STEWART, A. et al, (1980) *Social Stratification and Occupations*, London Mac millan.

WAINWRIGHT, J., FRANCIS, A. (1985), *Office Automation, Organisation and the Nature of Work*, Aldershot, Gower.

WEIR, M. (1977), *Are computer systems and humanised work compatible* ?, in Ottoway, R. (ed.) *Humanising the Workplace*, London, Croom Helm.

WERNEKE, D. (1983), *Microelectronics and Office Jobs*, Geneva, ILO.

WILLIAMS, R. & STEWARD, F. (1986), "*New technology agreements : an assessment*, "Industrial Relations Journal, vol. 16, no. 3.

WOOD, S. (ed.) (1982), *The Degradation of Work*, London, Hutchinson.

3

Technology and Traditionalism : An Indian Case Study

Suresh Kumar*

Industrialization is fast becoming a global process today sweeping each society into structural change. Particularly among Western scholars researches on the social aspects of industrialization and economic development have been the focal points of study in the past three decades. These studies undoubtedly present rich empirical data in the field of sociology yet they still lack the power for theoretical generalizations. The major reason behind this is perhaps the fact that the influence of industrialization on social structure is too complex. The social consequences of industrialization vary from society to society since this process normally happens to correspond with socio-cultural matrix of the particular society. It is not necessary, therefore, that the same significant consequences of industrialization experienced by one society should attend another industrializing society in the like manner. That is why in Indian context the findings and conclusions drawn by the Western social scientists although help us in many ways yet they hardly correspond with Indian

*Reader in Sociology, Ravi Shankar University, Raipur (M.P.).

socio-cultural matrix. In this reference it will not be far from truth to state that India still lacks in substantial work in this field. Now systematic investigations into the phenomenon of change in Indian context caused particularly by industry are urgently called far.

India particularly after independence has been experiencing the impact of considerable amount of economic development and industrial activity. The expanding network of various industries of different types and forms in almost every part of the country tends to bring about a tremendous change in the people's way of living, doing and thinking. Consequently the traditional norms and values of Indian social system also tend to come into the process of change. Such an organizational change proves to be one of the most important phenomena of major sociological significance, especially in respect of rural society which stands sharply exposed to the vital forces of the process of industrialization.

Chhattisgarh, particularly the tract around Bhilai in the present situation is indeed a fruitful ground for social investigation. But for the socio-economic survey conducted by the Directorate of Economic and Statistics, Madhya Pradesh, Bhopal[1] during the years 1955-59, i.e. before the commencement of 'Bhilai Steel Plant', the whole region remains more or less unexplored.

A thin edge of modern industrialism began to appear in this area only a decade after independence, i. e. towards the end of 1956 with the large establishment of Bhilai Steel Plant, the biggest steel works in the country. This huge establishment of steel plant with initial capacity of 2.5 million tonnes was undoubtedly an epoch-making event in the history of Chhattisgarh which for long had remained as one of the most backward and underdeveloped regions of India. Thus Chhattisgarh once quite an obscure region, arose from its age-long slumber and attracted almost international attention as the nerve centre of industrial activity of gigantic proportions. Initially the steel project acquired about 28,667 acres of land

1. Directorate of Economic and Statistics, M.P. Bhopal, *Socio-Economic Survey of Bhilai Region*, Vol. I, II and III, Bombay : Popular Prakashan, 1968.

(8,147 acres of Government land and 20,520 acres of private land) around Bhilai, a remote village in Durg district. The construction work began in the year 1957 with technical as well as economic collaboration of the Soviet Union. The production of the plant started in the year 1961. The expansion of the plant up to 2.5 million tonnes stage was completed in the year 1967. The field work of the present study was conducted just after the completion of this stage. During the years 1967-68 the main units of this steel plant comprised six Coke Oven Batteries, five Blast Furnaces, ten Open Hearth Furnaces and five different mills. Over fifty thousand workers were working in this huge organization.[2]

On the whole, the emergence of such a huge industrial organization substantially shook up the entire structure of the community. The entire tract particularly around Bhilai has been witnessing a number of various waves of change since the construction period of Bhilai Steel Plant. The following are some of the significant changes :

1. For such a large industrial sector not only a greater part of agricultural land was acquired by Bhilai Steel Project but the plant also occupied about eighty village settlements. All of these eighty villages were vacated. This event totally perplexed the rural community. Although they were successful in getting sufficient compensation for this but they had experienced a new state of being landless as well as homeless in their own native place. Ultimately the major portion of this population took shelter in nearby villages and settled themselves in neighbouring villages. Thus, a great amount of geographical movement took place in this area.
2. The remaining nearby villages were those who could save their dwellings but not the agricultural fields. They were also deprived of their greater parts of agricultural land. Labouring under such a drastic and severe acquisition of agricultural land, the

2. H.S.L. Bulletin, Public Relations Department, Bhilai Steel Plant, Feb. 70.

agricultural activity at once got reduced and impaired in most of these villages. In some of the villages the agricultural activity vanished altogether for want of sufficient agricultural land.

3. Immediately after the aquisition of land the construction work of this huge factory started in a large scale. In this situation it needed a large labour force. The project in this period made a provision especially for the villagers who had been deprived of their dwellings and agricultural fields and were completely out of job that the project will provide them with proper employment in case they would like to work in the project. As a result all of these villagers but for a few joined Bhilai Steel Plant. In this way this situation resulted into a higher rate of occupational mobility in this area.
4. A great amount of immigration also took place as the industry employed thousands of technicians, white-collar workers, skilled and unskilled labourers, etc. who came from almost every part of the country. This immigration also included those businessmen and traders who came here in search of their fortune and settled themselves in this region. As a result the rural community of this tract came closely in contact with various cultures, sub-cultures, languages, customs, etc. of the country.
5. The planned township of Bhilai Nagar which came into being near the plant beside the National Highway between plant site and Durg town emerged as the most attractive industrial urban centre for the villages within its hinterland. This neat and clean as well as well-planned town of Bhilai Nagar became very soon a nerve centre of trade, education, entertainment, politics and medical facilities for quite a large rural population around it. This has played a major role in bringing the ruralites in contact with urban industrial way of life.

The changes as discussed above seem to have shaken up the entire institutional structure of the region. The emerging industrial system which tends to supplant the traditional way of life has opened up for the people in this tract new fields of activity in which for proper participation they are becoming more receptive as well as adaptive in order to make themselves 'fit' to respond to new ways of life.

In every such society where industrialization exerts all its compelling forces a man's occupation becomes all at once an efficient indicator of his status and prestige because occupational placements in industrial system are always a matter of achievement, success and ability. Such occupational changes create changed images of status in the minds of the individuals and consequently such status mobility becomes very real phenomenon of crucial sociological significance. Status changes are bound to change the entire scheme of social stratification with all its ramifications and implications of such changes upon the very social structure. That is why social mobility is considered to be the dynamic aspect of social stratification itself.

It is obvious that the villages in the proximity of Bhilai Steel Plant are no exception to such inescapable forces of industrialization. It would therefore not be unsafe to assume that these 'impact' villages are certainly undergoing the process of transformation. Hence, it was considered worthwhile that these villages which are at present in the transitional stage, would offer themselves as wonderful laboratories for testing hypotheses related with the problematics of industry—induced social change.

The group of all the four villages taken up for intensive field research are in the closest proximity of the steel plant. The villages selected for the purpose are Khamaria, Katulbod, Somni and Kureta. The major consideration governing the selection of these proximate villages was to take up for investigation only those villages where the impact of industry should have been felt by the peasantry at the maximum, that is to say, which stand directly exposed to the force of industrialization and where industry is supposed to have made a real dent into the traditional way of life and living.

Since the objective of this investigation was to study the patterns of occupational change in industrializing set-up, each family head from each of these villages was thoroughly interviewed. Some of the joint families had more than one adult earners. These were also included in our list of the respondents taking into consideration the fact that the responses of such persons might also be equally valuable.

All the four villages are totally Hindu villages. The total population of Khamaria, Katulbod, Somni and Kureta is 2,542. Among these villages Khamaria has 180 families, Katulbod has 152 families and Somni and Kureta have 94 and 78 families respectively. It is a fact worth mentioning in this reference that there is a majority of nuclear families in all the four villages. Out of all the 504 families 425 families are nuclear whereas the number of joint families is only 79.

This group of four villages consists of 18 castes groups, namely, Brahmin, Kurmi, Teli, Rawat, Lodhi, Chauhan, Marar, Dhobi, Lohar, Nai, Kosta, Oria, Panka, Kewat, Dhimar, Gond, Satnami and Mehar.

Out of the total population of the four villages 76.2% are illiterate, 17.6% have studied up to primary school, 4.3% have done matriculation, 1.4% graduation and 5% post graduation.[4]

The Findings

The most significant and direct association between modes of economic production and social structure depends mainly upon the labour force participation and the economic activity of a particular society. As the economy of any society shifts from decentralized production system to the centralised and interdependent system, there is a series of change and shift in work roles. The ultimate result is that the whole occupational structure is changed. Opening of any large-scale industry is immediately followed by the initial formation of labour market and its rapid development attracts and covers the surrounding population for proper supply of its growing demand for labour force. In Indian context this may be regarded as the most significant aspect of industrialization, because in India it always happens at the cost of traditional modes of subsistence production system of the rural communities. The surrounding villages of Bhilai seem to have fully experienced this transformation in occupational structure.

In what follows the changes in the occupational structure of the villages have been analysed. For this purpose the relevant information has been collected by interviewing each case regarding the means of his livelihood before and after the opening of Bhilai Steel Plant. 30.24% of our respondents were not available for employment at the time factory was started in the area. They are still below the age of twenty five. Such cases were no pursued in analysis of occupational change caused by industry. However, 69.76% cases had already embarked upon their career and hence an enquiry could furnish details on career shifts after the establishment of Bhilai Steel Plant. Only these cases have been taken up for analysis in before and after situation to bring out clearly the role of industry accounting for occupational change.

An effort has been made to show the way industrialization has altered the occupational structure of the 'select' villages. The related findings reveal the following facts.

1. In all the four villages the respondents had greater tendency to pursuing agricultural occupations before the opening of Bhilai Steel Plant while after the establishment of the factory the percentage of the respondents following agricultural occupations came down considerably.

2. There is a noticeable fall in the percentage of artisans and servants as revealed by the data tabulated. Industrialization has caused many of them to switch over to alternative occupations.

3. There is a considerable increase in the number of petty traders and casual labourers after the opening of Bhilai Steel Plant.

4. Village administration and school teaching remain largely unaffected. One more village official has appeared on the scene. A village-level-worker has been posted at Kureta, representing a welfare arm of community development programme. Khamaria and Somni both have got one more school teacher.

5. Some of the villagers had the tendency to take jobs in urban areas before the establishment of Bhilai Steel Plant,

just for the fact such jobs would turn out to be more rewarding. In Khamaria three respondents have been put under this category. They are a Kewat, a Gond and a Kurmi. They did not follow any of the occupations available in the village. The Kewat worked as a hotel bearer in Durg and the Gond worked as a porter at bus-stand in the same town. Both of them are now working in Bhilai Steel Plant as manual workers. The third case has made numerous changes in occupational career. Ultimately he has fallen back upon agriculture and is presently enjoying the status of owner cultivator in Khamaria. At first, he worked as compounder in a dispensary in Durg. After a lapse of couple of years he started medical practice on his own in the same town. He did not care however to obtain the medical licence. He had to give it up for fear of legal action. Next, he laid his hands on salesmanship but there also he did not do well. At long last he came to realise that temperamentally he was not fit to serve anyone. And for some time he did not do any work. He simply went out of his village in search of some gainful employment. But he was not a success anywhere. At the death of his father he inherited say, fifty acres of cultivable land. He is a settled agriculturist now in his own village.

All the five respondents of Katulbod in this category are Kurmi by caste. This group included a doctor and three engineers working at various places. They have now joined B.S.P. services. The fifth man was a clerk in Durg collectorate. He has retired from service and now is settled as a grain merchant in Katulbod. A Kurmi of Somni who formerly worked as a peon in a Government office at Raipur has now returned home and is settled now as a tenant farmer. The two respondents of Kureta, a Teli and an Oria formerly worked as a peon and a bus cleaner at Durg. Both of them are presently manual workers in Bhilai Steel Plant.

It is to be observed that many of those who had gone at various jobs, have come back to their native place in the wake of industrialization. It is further to be noted that a good majority of such respondents has joined the services at Bhilai Steel Plant.

6. The findings further indicate that a considerable

majority of the respondents prefer to work in Bhilai Steel Plant. Such respondents constitute nearly one third of the total number of cases of our universe.

7. Findings corroborate that the occupational structure of industrialising society is repatterning itself. It is a gradual switch over from agriculture-based to industry-base.

Analysis of the data presented above abundantly reveals that the establishment of the huge steel plant has altered substantially the entire occupational structure of the adjoining villages. In what follows an attempt has been made to explore the patterns of occupational mobility. For the purpose, each case of occupational change was taken into account in order to analyse in details the way it took place.

1. From amongst those who were owner cultivators before industrialization 40% have moved over to other occupation. A significant factor responsible for such a drastic switch over was that cultivable land possessed by these owner cultivators was acquired by the B.S.P. Save 5.2% who still opted to become tenant farmers in their villages, all of such cases gave up agriculture altogether. 11.2% started trading on a petty scale and the remainder were lucky enough to have found jobs in B.S P.

2. Because of scarcity of agricultural land only 42.9% of the tenant farmers could hold on to their occupation. Nearly half of those cases who switched over to alternative occupations joined the factory. Nearly eleven per cent became petty traders. The remainder turned out to be casual labourers.

3. More than half of those who were agricultural labourers before the wake of industrialization in this area are now in pursuit of other occupations. Among them, only 20.9% could get B.S.P. jobs and 4.5% engaged themselves in petty trading. The remaining 25.2% of them either work as domestic servants in their villages or work as casual labourers in the industrial area.

4. From amongst those who were traditionally skilled professionals, 29.5% of the cases changed their occupation, as if, they were wise enough to feel well in advance the unsuitability of their traditional professions in the emerging

industrial complex. However, only 20% of them who either took to B.S.P. work or started petty trading seem to have changed positively for the better. The remainder who had to became either domestic servants or agricultural labourers after leaving their traditional occupation are economically depressed.

5. Because of the fact that a casual labour in industrial area earns attractive wages, 25% of the individuals previously engaged either as domestic servants or as graziers in the villages have now become casual labourers. Besides this, 11.5% now have joined B.S.P. and think themselves fortunate enough so as to have found work with the industrial organisations. A small number is doing petty trading and other odd jobs.

6. Before the opening of the factory the number of the casual labourers in the villages was very small. During the construction work of the factory every one of them tried one's best to find a job in the factory, but only 50% of them could succeed. The remainder still depend on casual labour. The point worth noticing here is that all of such cases feel that they are in a better condition than before.

7. No case of mobility was found among those who were formerly occupied either as 'Kotwars' or as school teachers. They remain as before in a 'no change' position.

The Role of Caste

India is a caste ridden society. It regulates every aspect of a Hindu's life in India. In a caste system admittance to a particular social stratum is determined by birth. Birth into a caste, in theory, determines what food to eat, with whom to have physical contact, how to conduct, what occupation to follow, circumscribes the choice of marriage partner as well as determines the ways of acting and thinking.[5]

One of the most apparent features of caste with which it is generally characterised is that it is closely associated with occupations. Many castes are often named in terms of occupations, e.g. *Lohar* (blacksmith), *Sunar* (goldsmith), *Teli* (oil presser), etc. In fact, caste and occupation have always been put together mainly because they are traditionally associated since long. Each caste has a particular occupation which is considered to be its traditional occupation. According to Ghurye it was believed that every one of the large number of castes had an occupation which was its own, its traditional linkage and hence the hereditary occupation of its members to which search of another was at least not proper, if actually not sinful.[6]

Most of the recent scholars have also recognised this association between caste and occupation. Majumdar accepts that it is the only basis of the system of social stratification in India.[7] Srinivas,[8] Dube[9] and Mayer[10] have also shown that caste and the pursuit of occupation are closely linked.

The Perception of Traditional Occupation

Ideally in Hindu Society occupation is also a matter of ascription. Members of a caste are expected to hold to their traditional callings. Theoretically, this is the only correct and proper way for them to make a living. In practice, however, persistence to one's own traditional occupation is still seen in higher percentages in rural communities in comparison to those of industrial and urban communities in India. This has been proved by some of the recent studies made in this regard.[11]

Despite various recent findings in this concern, there still exists the problem of finding a valid and a proper approach to study this particular phenomenon on the basis of which one could precisely judge the cases of conformity or non-conformity to the traditional occupations. The most confusing aspect of this is that the most of the researchers of caste and occupation apply their own discretion in deciding the traditional occupation of a particular caste. Such criteria may be lebelled as objective, but there is every possibility of arriving at biased and dubious conclusions in such efforts. The notions about the traditional occupations of caste groups have been changing from time to time. Also the idea about traditional callings in a particular caste group may vary even from one region to another at the same time.

It is for the reasons mentioned above that a subjective

approach has been adopted in this study to judge whether or not a respondent holds to his traditional occupation. Thus, before arriving at any such conclusion the efforts have been made in the first instance to know the subjective perception of the respondents about the traditional occupation of their own caste groups. On the basis of findings, the cases of conformity or non-conformity to the traditional occupations have been judged.

In order to make things a little more concrete an open question was put: "What is your traditional occupation ?" The above question was put to the respondents of all the eighteen castes of all the four villages. In spite of large consensus among the respondents a little variation was found in the perception of their caste-callings.

The difference in perception does not mean here that each caste member has altogether different perception about his traditional occupation. In fact, the difference lies only in the point that in spite of agreement with the common perception some of the respondents had a tendency of adding something to it. For example, 62.5% of Brahmins perceived their traditional occupation to be the 'priesthood' while the rest of them were of the opinion that priesthood alone should not be taken as the traditional occupation of the Brahmin. Along with priesthood agriculture also should be taken as their traditional occupation, as the same is also being pursued by their forefathers since long in this region.

A brief analysis of the findings in this regard may be presented in the following manner.

1. First and foremost, there was obtainable a large consensus among the respondents over their traditional occupations.

2. There existed a tendency to perceive more than one traditional occupations.

3. However, it is to be noted that the percentage of the respondents with perception of only one occupation is much higher than those who speak of two occupations.

4. Eight caste groups had full consensus about their traditional occupations. Those castes were Kurmi, Lodhi, Marar, Dhobi, Lohar, Nai, Kosta and Mehar.

5. In a face-to-face situation it was felt that each respondent had a clear perception in this regard and not a single case of indecision was found. Rather, an immediate response was received in almost all the cases.

On the basis of the above findings all the occupations perceived by the members of a certain caste as their traditional occupations were regarded as their 'traditional occupations' in matters of judging conformity or non-conformity. This effort may be significant also in exploring the extent to which a respondent is sincere about following the occupational norms of one's caste.

It brings out that a good majority of the respondents did not follow its traditional occupations. The percentage of the cases of conformity is only 38.14% while the cases of those who are not following their traditional occupations constitute 61.86% of the total respondents. It shows that the changed structure of economy has affected a great deal the traditional norms of caste and occupational pursuit.

It can also be observed that our respondents knew very well that they had departed from their caste-based occupations, because in spite of their having a good perception of their own caste-callings they had deviated from them. According to the views of many of the cases the conformity to one's own caste-calling is getting difficult in the economy of the present situation, especially so, in the expanding complex. For example, big oil pressing industries and steel foundries have come to be established in the adjoining industrial area and as a result thereof oils and agricultural hardware tools have become available in plenty and at much lower cost in the village markets. So most of the traditional oilpressers and village blacksmiths found it uneconomical to carry on their traditional calling and under compelling circumstances have given it up. Same is the case with many who should even like to stick to a traditional calling. The opening of laundries and dry cleaning and washing companies in the nearby industrial area has belittled the importance of village Dhobis. The modern hair cutting saloons in the close by have taken away all the custom from the hands of village barbers. Demand on the Brahmin to function as a priest is on the decline with declining faith in religion and increasing secularization in attitudes towards life. Some of the castes i.e. Kurmi, Lodhi and Oria who were solely occupied in agriculture had to take to alternative occupations in face of paucity of cultivable land.

Thus, industrialization has exerted a great 'pull' and is responsible for bringing about change in the pattern of work in such a large measure in the complex of this region. The huge steel industry attracted these villagers and inspired them to leave their less profitable traditional occupations. Though holding to one's own traditional occupation is an ideal for every individual in Hindu social system, yet economic considerations, rising level of aspirations, unsuitability of an occupation in changing circumstances and new openings and temptations thrown up by industry has taken individual on to a path of non-conformity. In this way industrialization has revolutionalized the traditional occupational structure of the villages of this backward region.

The most significant aspect of such occupational change is that it is likely to affect relations between those who are 'high' and 'low' in society. A persistent theme running through the literature on the sociology of work concerns the notion that a man's occupation exerts a compelling influence on his behaviour, both off and on the job. Knowing thus a person's occupation, it is argued, enables us to make a number of reasonably accurate predictions about his attitudes, life-style and central values. Accordingly, a change in occupation, which is one of the chief indicators of an individual's status, in many ways prepares a person's status mobility. Specially in situations like those of industrializing villages of Khamaria, Katulbod, Somni and Kureta, where the rate of change has gone considerably so high, ascription-oriented values are likely to lose their importance; and the emerging system of social differentiation, mostly based on new patterns, becomes more mobile, more open as well as more complex than before.

REFERENCES

1. Directorate of Economics and Statistics, M.P. Bhopal, Socio-economic Survey of Bhilai Region, Vol. I, II and III, Bombay, Popular Prakashan, 1968.
2. H.S.L. BULLETIN (1979), Public Relations Department, Bhilai Steel Plant.
3. SEYMOUR M. LIPSET AND REIRHARD BENDIX (1963), *Social Mobility in Industrial Society*, University of California Press, pp. 1-2.
4. SURESH KUMAR (1986), "*Social Mobility in Industrializing Society*", Jaipur, Rawat Publications.
5. CAROL OWEN (1968), "*Social Stratification*", London, Routledge and Kegan Paul, p. 14.
6. G.S. GHURYE (1961), *Caste, Class and Occupation*, Bombay, Popular Book Dept. p. 241.
7. D.N. MAJUMDAR (1962) *Caste, and Communication in an Indian Village*, Bombay, Agra Publishing House.
8. M.N. SRINIVAS (1962), *Caste in Modern India and other Essays*, Bombay : Asia Publishing House.
9. S.C. DUBE (1965). *Indian Village*, London : Routledge and Kegan Paul.
10. ADRIAN C. MAYER (1965), *Caste and Kinship in Modern India*, London Routledge and Kegan Paul.
11. K.S. MATHUR (1964), *Caste and Rituals in a Malva Village*, Bombay : Asia Publishing House.

4

Forecasting and Researching New Technology in the European Community

Werner Wobbe*

The changes underway, partly caused by scientific and technical progress, make it difficult to obtain a clear picture of what European societies could and should become in fifteen to twenty years' time. Consequently, there is an ever present need for European forecasting in science and technology to clarify and give a common purpose to Europe's long-term technical and socio-economic development. This is all the more important today when Europe is suffering from the lack of any integrated view of national futures and when science and technology are confronted on a planetary scale with new industrial, military and political challenges.

Central questions for the future are : How can the technological "renaissance" of Europe be promoted ? How can we cope with the changes in employment brought about by the new technologies ? On which actors can new policies be based, given the increasing world-wide dimensions of science, technology and industrial structure implying also world based strategies ? How can we improve European living conditions (health,

*Head of research sub-programme 'Technology, Work and Employment' (TWE) within the FAST programme, Brussels, Belgium.

nutrition, environment, participation) ? How can European science and technology also be harnessed for the benefit of the Third World countries which will account of four-fifths of the world's population in fifteen years' time ?

These problems are tackled by the FAST Programme (Forecasting and Assessment is Science and Technology). It is the second programme of its kind within the Commission of the European Communities (C.E.C.). FAST I commenced in 1978 and FAST II will terminate in 1987.

FAST I (1978-1983)

Societal analysis and forecasting of technological change may seem a circuitous way of identifying long-term R & D priorities but it is unavoidable. This was the basis of the FAST approach[1] ; its work, backed up by studies conducted by some 50 European centres, showed that the Community's R & D policy, inspired by the crisis of the 1970s did not sufficiently anticipate the turn of the century problems.

Today, this principle is increasingly prompting national authorities to take measures in this field. Examples include the campaign launched in Federal Germany by the Federal Ministry for Research and Technology on 'Sozialwissenschaftliche Technikforschung', the new five year research programme entitled "The Context of Change" of the Economic and Social Research Council in the United Kingdom and the major programme recently approved in Denmark on technology and society.

The FAST approach also proved fruitful in practical terms ; the results of the approach have given rise to new joint ventures. For example, thay have been used to prepare a Community biotechnology policy ; they have provided initial concepts for the ESPRIT programme ; they have provoked a debate on the role of technological innovation in the phenomenon of

1. This work, conducted around three priority themes in the programme (work and employment, the information society, biotechnology and society), has yielded around 100 research reports distributed throughout the scientific community and is still in wide demand.

See the final FAST I report, Eurofutures—The Challenges of Innovation. Commission of the European Communities, Butterworth, London, 1984. A French edition is also available : "Europe 1995", Mutations technologiques et enjeux sociaux"-Futuribles, Paris, 1983.

"bogged-down growth." In a more limited area, they have, for example, led to national programmes in the areas of chemical R & D and instrumentation technologies for the environment, etc.

FAST II (1984-1987)

The main task of FAST is still to analyse scientific and technical change in its many dimensions—economic, social and political—so as to identify new priorities for a common R & D policy and for long term action in the Community.

In addition, FAST aims to strengthen the bases of European forecasting and forward thinking by exploiting the research in the long term undertaken in the Member States and encouraging the setting-up of *ad hoc* cooperation networks between researchers and potential users.

Present FAST II research activities are focused on five problem areas which are crucial for economic growth and social development :

**Service Activities and the New Technologies* (SERV Programme)
Analysis of future developments in service activities around the new technologies, highlighting of the lines to be followed in Community policies to ensure that Europe can benefit to the maximum from this change in order to adopt and improve its production system.

**The New Strategic Industrial System of Communication* (COM Programme)
Identification of the long-term economic, industrial and social issues involved in the development of the "communication function" and its mastery by European society.

**The Future of a Food System* (ALIM Programme)
Analysis of current and future trends in the European food system as regards both its agro-industrial aspect and its relations with the consumer and his health.

**Integrated Development of Renewable Natural Resources* (RES Programme)
Analysis of the prospects offered by recent scientific and

technological breakthroughs, especially in environmental monitoring, modelling and biotechnology, with a view to the coherent management of the system of natural resources in Europe and the integration of the relevant policies.

**The Relationships between Technology, Work and Employment* (TWE Programme)
Analysis of existing options for the utilisation of new technologies with a view to the management of employment and reorganisation of work better geared to the Community's economic and social objectives.

Each of these five sub-programmes is based on a series of research activities (contract research, conferences, network activities, etc.). They are described in documents similar to this one.

The FAST Programme has established and developed an European system of cooperation designed to contribute to the emergence of an European concept and application of prospective research.

The system is based upon :

—70 *research teams*, under contract to carry out a large part of the research activities ;

—"12+1" *network*, a research and communication tool, linking FAST with ten national research units designated by the Member States. These units play a role in their respective countries in the analysis of the long term implications of technical change ;

—Informal "*ad-hoc*" networks set up around individual research activities, and bringing together experts from academic, socio-professional and public sectors.

—*Scientific fellows* seconded to the FAST programme by the Member States.

The programme has a total budget of 8.5 million ECU (equal to approx. US $ 1 million) for the period 1983-1987.

A more detailed description is given below of the Technology, Work and Employment sub-programme (TWE programme) within the FAST research framework—the questions it poses, research areas and expected results.

The TWE Sub-programme : Technology, Work and Employment

In Europe, *employment* is likely to continue to be a major problem throughout the 1980s, one of the intervening factors being the technological changes brought about mainly by the new automation, information and communications technologies. The tidal wave of *process innovations* that has radically changed the production apparatus in the advanced industrialised countries over the past twenty years will sweep on in the form of the increasing dissemination of *new advanced production equipment* (robotics, flexible manufacturing systems, etc.) and the slow but sure penetration of the new biotechnologies and new materials. This will speed up the *processes whereby human work is replaced by the machine* as a source of mechanical energy and/or manager of repetitive and routine information, consequently of a low added value.

The future of employment does not depend solely on technology and its use, however. As employment is one of the essential dimensions of *work*, it is clear that a "long term" solution to unemployment should be found by *innovation in work*, amongst other things by *separating the ideas of work and paid employment* (thereby bringing about a new relationship between income and work) and by promoting new man-machine relationships.

What type(s) of innovation(s) then ? What *European work scenarios* ? What will the future hold, for example for the "brainworkers", the "farmworkers," and the "Over-Fifties" ?

What will be the scope and significance of distance working over the next 10 or 15 years ?

The *employment issue* arises mainly in *manufacturing industries* (mature and/or expanding) and *services* : what will then be the implications and consequences for Europe and its scientific and technological policy of the new *integrated flexible manufacturing systems*, the *new materials* and the "*technologies of light*" ?

These are the questions to be tackled by the TWE programme in its nine research activities that should provide the raw material for five "products,"

*European scenarios for work
*Four strategic dossiers, i.e.

—*Man-machine relationships* (DOS 1), the goal of which is to identify and set up hierarchy for the technical, economic and social challenges of the increasing automation of human activities in order to suggest technological innovation policies centering on man rather than the machine.
—*Manufacturing sectors in turbulence* (DOS 2), the aim of which is to identify the development options open on a European scale for manufacturing industries in turbulence that will remain of major importance for growth and employment.
—*Transnational and local technologies* (NOV 1), where an attempt to identify the new relationships between the dynamics of the transnationalisation of technological and economic systems on a world-wide scale on the one hand and the processes of taking advantages of local resources and local uses of the new products and services associated with technological change, on the other and.
—*The self-employed* (NOV 2), the aim being to investigate the possible scope and practical significance in European societies of new forms of "independent" patterns of employment. It would consist of an in-depth study on one aspect of the work being carried out under the heading of European scenarios for work.

To establish the above mentioned dossiers, the FAST team is supported by a broad range of European research teams. The studies are grouped around nine themes.

—Scenarios for Work in Europe.
—The End of the Farmworkers—New Farmworkers
—Teachers : Transformations and Prospects
—Brainworkers : Typology, Training Background and Work Situation.
—The "Over Fifties"
—Distance Working in Urban and Rural Regions of Europe.
—New Technical Production Systems : Work in the Factory of the Future.
—Industrial Prospects for the "Technologies of Light".
—Industry in Transformation : The New Materials.

5

Automation in Small and Medium-Sized Firms

Peter M. Van der STAAL*

1. Introduction :

In the Netherlands a widespread expectation exists that advanced automation systems will have a radical impact on the structure of medium and small-sized firms (MSFs) and on their relations with larger enterprises. These influences can be exerted indirectly by changes in the external relations as a consequence of automation in the relevant environments of MSFs. More immediate effects are expected by automation of internal production and information processes.

Regarding their own process innovation, MSFs possess advantages as well as drawbacks in comparison with larger enterprises. Usually bigger companies have more finance capital at their disposal, employ their own automation experts and are capable of running higher risks.

On the other hand, smaller companies may have a larger flexibility in changing from products, processes, jobs and organizations. Moreover predominantly small and medium series are being produced in MSFs. For these reasons advanced flexible automation techniques seem to be more promising

* Leiden University, BEWETON Futures Research.

for MSFs than the earlier forms of fixed automation. Larger companies are technically, financially and scale-economically more committed to fixed automation systems. The perceived predisposition of MSFs for flexible automation might lead to an accelerated catch up process of these firms.

In this study one of the central questions is whether, up to what extent and for which sectors these expectations are realistic. Special attention is given to aspects of flexibility, standardization and centralization. Related questions concern the effects of automation and positions in the production structure.

2. CHARACTERIZATION OF MEDIUM AND SMALL SIZED-FIRMS

In the Dutch statistical sources small and medium-sized firms are defined by the numerical criterion of 100 employees. MSFs play an important role in the Dutch economy. In 1982 they represented 72% of the total number of enterprises. The share of the MSFs of the National Income is 39%. The employment amount to 43% of the labour force. Consequently the productivity in MSFs is slightly less than the average. This is partly explained by the over-all shift of labour volume of the service sector in which advantages of scale play a less important role than in industry. Within the MSFs almost the same sectoral shifts occurred as in the over-all economy. Crafts and small industries arc structurally under pressure. Retail trade shows a slight decrease. Wholesale trade, banking and insurance and the private service sector are growing steadily. Some relevant aspects of these sectors will be dealt with next.

2.1 Industrial crafts and small industries

These enterprises are active in the manufacturing, metal- and construction industry, reparation services and the like. MSFs are underrepresented in the electrotechnical and petro-chemical industry and to a lesser extent in textile and means of transport industries. They are overrepresented in wood and furniture-industry, in clothing, leather and shoe-industry and in instruments and optical industry and to a lesser extent in construction material-pottery-and glass-industry and the metal-industry. MSFs are especially important in sectors where discrete products are made in small series, in subcontracting and in crafts. They play no role in the process-industry and in mass manufacturing and assembly. The over-all labour productivity is lower but the production process is generally more flexible in smaller firms, which is of special significance in the production of small series.

2.2 Retail trade

In the sector of retail trade 96% of the firms employ less than 10 persons and 0.2% are large stores. Lower household budgets and intensified competition tend to decrease the market share of small firms compared with large stores. In retail trade about 30% of small stores are committed to commercial cooperations as multiple shop organizations, a franchise chain or a purchase combination. In practice this form of cooperation is also leading to initiatives and cooperation in the implementation of shop automation systems.

The food sector is mainly characterized by a relatively large number of standard articles, small profits and quick returns. The turnover of small firms is absolutely and relatively decreasing steadily, due to demographic shifts and consumption levels and patterns and to increased competition with large stores. Large supermarket chains tend to put its stamp on this sector more and more.

In the non-food sector the turnover is smaller, the profits are higher and the assortment is more differentiated. A number of branches is sensitive to fashion of season. The service function in the specialized small stores is very important. The market situation and perspective for the MSFs is developing even worse than in the food sector.

In stores with a production function such as butchers and bakers a tendency to concentration and increase in scales can be observed. There is even a growing tendency to displace sales to separate general food stores.

2.3 The service sector

This sector offers a wide variety of services. In banking

and insurance larger enterprises can be found. In branches such as brokers, in-house architects, fashion designers large enterprises do not exist. The sector as a whole is highly dynamic and is steadily growing.

The banking sector is characterized by a limited number of large banks and a large number of one man or small firms. In the insurance sector a similar pattern exists. The number of independent intermediating persons is rising steadily. The insurance turnover is slightly decreasing.

In the private service sector the large companies can be found mainly in the accountancy and in book-keeping; in architecture and engineering consultancy and in personnel recruiting firms. The technical consultants suffer from decreasing turnover. The other part of the service sector is profiting from the economic revival, not in the least because of the early implementation of automation.

3. ADVANCED AUTOMATION TECHNOLOGY

3.1 Electronic automation

Automation is here defined as the decreasing human intervention and the increasing use of machines in the steering, control and coordination of processes for the production of goods and services. The considered automation techniques are based on micro-electronics and computer technology. The new dimension of electronic automation is the programmability. The electronified system can perform an indefinite large variety of tasks. This fact enables the same machine to relatively easily switch to another production process. This possibility fits well the increased need of diversity and variety of products. Programmable automation can be applied in various production processes. As to primary processes the following forms can be distinguished :

1. Manufacturing automation.
2. Administrative automation.
3. Transaction automation.
4. Automation of information transmission.

These forms will briefly be dealt with next.

3.2 Forms of automation

3.2.1 Manufacturing automation

According to general expectations, electronic production automation will have a major impact on the manufacturing of small and medium series of discrete products. Whereas now the natural flexibility of human operators of functionally specialized machines save time and costs in switching to other products, in the near future this flexibility will be built in the machine systems. This possibility emerges with the combination of computers and production systems. Computers are being used to process production data, to store and retrieve these data and to communicate with machines or machine systems. The three main functions in the discrete manufacturing are the generation of product information, the actual fabrication and the control of this process. The computer-machine systems are illuminated per main function :

a. *Computer aided design* : Computer aided design systems help design engineers and draftsmen to elaborate their ideas of new products. With drawing elements and components, stored in the computer memory, a new product concept with variants can be generated, be displayed on a screen, be calculated and be iteratively developed into a new product. The productivity of most of these activities can be increased substantially. It becomes possible to satisfy quickly, repeatedly and flexibly changing specifications of customers. The electronically stored product specifications can be coupled with the machine control systems. This offers an enormous potential to increase efficiency, information and process control. It can eliminate work in the production planning.

b. *Computer aided manufacturing* : With this term is indicated the programmable automation on the shop floor. The most simple form of computer control is a stand alone computerized numerically controlled (CNC) machine tool, equipped with a dedicated minicomputer and a screen and keyboard. This allows the operator to write or edit a programme at the machine. For reasons of efficiency the programme may be prepared off-line by a programmer. In the case of direct numerical control (CNC) more than one machine tool is programmed and controlled by a central mini-or mainframe computer. Intelligence may also be distributed by equipping the machine tools with microcomputers. This makes operation less vulnerable and even more flexible.

A further development is the flexible manufacturing cell (FMC), a cluster of cooperating machine tools and robots, which are coordinated by a computer system. If these cells are coupled by automated materials handling systems we have a flexible manufacturing system (FMS). This system is still out of reach of MSFs though.

c. *Computer aided management* : For assisting management in the control of a manufacturing operation, computer systems are developed to coordinate and manipulate databases containing current information about the main aspects of the manufacturing process.

Management information systems (MIS) collect and process data on designs, inventory, orders, personnel and machine capacities, etc. and summarize them into corporate information for decision making by the management.

Computer aided planning systems such as materials requirement planning (MRP) and Computer aided process planning (CAPP) assists the factory management in scheduling the work flow and to optimize decisions and operations sequences.

3.2.2 Administrative automation

Administration can be defined as activities in which information is processed which has no direct function in the real production process. Because, in administration, information itself is material, workpiece, endproduct and distribution object, computer systems can be used there more directly and efficiently then in material production processes. For textual data electronic writers and wordprocessors now become widely available. Standard software can be bought and, sometimes after slight modification, be applied for business administration and planning, financial and commercial administration, wages and salaries administration and project and production administration. The possibility of the coupling of the new generation of office automation systems will have a radical impact on information processes. An integration of separate systems becomes possible by the connection through telephone lines and data networks. Administrative information can be communicated within the different administrative functions as with the production and distribution functions ; within an organization via local networks or between companies *via* wider area networks. Electronic data traffic accelerates communication by the elimination of labour intensive and time consuming activities. Via terminals the extensive memories, software and processing capabilities of central mainframes can be utilized. As expected, productivity will also be raised on the management level in the administrations by the implementation of multifunctional workstations. Generation of management information and external communication will be accelerated and optimized.

3.2.3 Transaction automation

Transactions can be defined as the exchange of money, goods and services between consumers, trade firms and financial organizations. A distinction can be made between commercial and financial activities.

For the automation of commercial intercourse technical systems become available under the name of Electronic Funds Transfer Systems (EFTS). These systems consist of a terminal, attached to a cash register and connected with a microcomputer built in a master cash register or with a central computer. An article number should be keyed in on the terminal or read by an electronic laserscanner from a bar code, attached to the articles and containing article information. The computer automatically searches the corresponding article data and price, which is added and transmitted to and printed on the cash register. These systems are indicated with the name Point of Sale (POS). The information can also be coupled to a central administrative or to a wholesale trade computer, to help the management with inventory management, administration and ordering.

The article and price data can also be coupled with Payment functions in Point of Payment Systems (POP). The coupling is made *via* a telephone or a dataline with financial institutions, with the help of a magnetic or electronic card, containing information, on a personal identification number, the bank relation, the account number and a bank balance.

In financial institutions automation is already further developed, especially for the administrative execution of mass transactions on the central level.

In retail banking the most important functions are counter transactions and data input in the central administration. For automation of cash transactions a variety of cash dispenser systems are operational, with which clients can cash money or execute other transactions without the intervention of a cashier or other bank personnel and unrestricted to office hours. Via telecommunications systems even home banking becomes possible. Automation of financial traffic will carry through further if the various apparative applications as home terminals, POP systems; cashier and transaction systems are integrated *via* telephone and datacommunication lines and networks.

3.2.4 Automation of information transmission

In the domain of transmission of information new developments emerge by the computerizing and digitalizing of information. This area is indicated by the term datacommunication.

Datacommunication is the exchange of digital information between two or more dataprocessing systems such as a computer, a microprocessor, a wordprocessor, a cashterminal, a machine control system, etc. Transmission channels may consist of a cable, a network, a telephone line or a satellite connection. For the communication between computers special networks are installed. Dependent of the distribution area they are named local or wide area networks. Digitalizing of networks and peripheral systems and the use of wide band communication channels [satellite and glass fibre] will enable a better integration of digital terminals of all kinds. The before mentioned transmission systems will cause a substantial increase of numbers and type of connections and forms of information. They open and extend the possibilities for the telecommunication of data, sound, videosignals, images, documents, text, financial transactions, letters, etc.

Datacommunication not merely adds something to the possibilities of automation. Specifically the superfast and automatic transmission of data between computerterminals is vital for the realization of the full potential of this automation. Massive financial and commercial transaction traffic will run faster, more flexibly, more reliably and more controlled. Connection to networks give smaller computer systems access to the facilities of other systems and the larger databanks, faster processing capacities and special applications programmes of large mainframe computers. This increases the availability and exchange speed of information drastically. Telecommunication forms a powerful synergetic combination with computertechnology.

4. DIFFUSION OF AUTOMATION IN MSFs

The described automation techniques are not all operational or widely diffused in MSFs. Some are still in a conceptual, developing or experimental state. Other techniques are still exclusively applied in early adopting firms. The systems which are already or will soon be operational will be described for three important sectors.

4.1. Automation in small industries

In small industrial firms computer automation is becoming operational in production and administration. Because usually they do not have a significant design and construction unit, computer aided design systems were not yet widely adopted in small production firms. The emerging software for designing and draughting, developed for microcomputers will bring CAD systems into reach of medium-sized firms. Up to the present CAD for production automation is primarily developed and applied by consulting engineers in the service sector. Automation of machine tools is rapidly introduced in medium-sized firms. This is mainly a process of a gradual substitution of obsolete conventional and, the already wider penetrated, numerically controlled machinery.Programmes are either written on the CNC machine or on a separate programming system or the facilities of a larger programming package of a computer service or engineering organization are being used *via* telephone lines.

Industrial robots, varying from simple pick and place machines to advanced handling and working machines with many degrees of freedom are also installed in medium-sized production firms. The main applications are simple welding tasks and paint spraying. Integration by means of more complex configurations of machines and departments is considered in MSFs, but still hardly realized as a consequence of high investments and short depreciation terms, of the vulnerability by dependency on technology and of the inherent problems of personnel and organization.

4.2 Automation in distribution

In distribution a distinction can be made between automation of transactions and of administration. Transaction activities can be subdivided into the transfer and the payment of goods.

Well known and widely diffused systems for the registration of the goods stream are the (master) cash systems provided with price look up computer, laser pen or-scanning. The largest application is in the food sector. A vast majority of foodstuffs is already provided with a bar code. Generally these systems are coupled with administrative computers.

The overall penetration of computers for commercial, inventory and financial administration is still low, especially in retail trade. With the exception of the food stores the same holds for ordering *via* terminals of the so-called order entry or order pick systems, A system is being developed for automatic communication between computer systems of factories, wholesale houses and retail trade organizations. Some experiments are carried out between factories, wholesale firms and distribution points for larger stores.

Coupling with financial institutions seems more promising than successful. Experiments with tele-shopping of consumers *via* interactive networks appeared to have little success. Electronic paying *via* POS and POP systems asks for a complex system computers terminals and networks. A large experiment with electronic cards, petrol stations is being conducted.

4.3 AUTOMATION IN THE SERVICE SECTOR

4.3.1 The technical service sector

Engineering consultants firms were relatively early in adopting Computer Aided Design systems based on minicomputer. The rising availabitity of microcomputer software will undoubtedly accelerate this trend further.

Architects appear to lag behind this trend. Only some larger and a few medium-sized bureaus have implemented a CAD system so far. CAD systems are being used for designing drafting, calculating, constructing, specifications and work-drawings. Other computer systems are used for specifications and estimates, hour calculations and financial administration.

4.3.2 Administrative and financial services

Automation offers the possibility to increase the diversity and speed of the traditional services in this branch and to offer new services. Accountants and organization consultants, for instance, appear to give advice in question of automation. The net effect of automation in this sector seems to develop positive on the whole.

In the accountancy and administration offices the degree of automation is very high. Plenty of software is available for activities such as book-keeping, debitors and creditors-administration, calculations of salaries, working hours, declarations, etc.. This sector mainly executes the administration for MSFs. Automation in this sector can also effect the automation of the MSF client system indirectly.

Automation in the insurance busineses is mainly in administration and communication. Word-processors are implemented for correspondence and manipulation of text fragments of policies and forms. For insurance consultants a central communication system exists for the exchange of information with insurance companies and private branch organizations. A videotex system is being developed for damage experts and repair shops for the standard calculations of repair costs. Automation is initiated by branch-organizations. The efforts are directed at standardization and integrated datanetworks. Individual small firms can exert little influence on these developments.

In banking two information systems can be distinguished : administration and financial traffic.

Automation on mainframes of the administration of massive streams of money on the central levels is already in the mature phase. The processing of the input data in local branch offices and affiliations is now on the verge of automation. Pioneering firms have started to automate the processing of vouchers, transfer checks, etc. In this phase paper information is converted to diskettes for off line processing in central computers. Automation of cash transactions in the Netherlands is lagging behind the surrounding countries. In a relatively small but increasing number of banks automatic cash dispensers are being installed. Tele-transactions *via* point of payment terminals, telephone lines or datanetworks are merely experimental.

4.4. The automation degree in MSFs

Although on the basis of the case studies generalization is not justifiable, there is a strong impression that the degree of automation in MSFs is still far lagging the automation in bigger firms and that the high expectations did not yet have been realized. Some recent quantitative material regarding this matter released by the Dutch Bureau of Statistics on the basis of a wider survey supports this impression. The following matrix indicates the automation degree by the end of 1985 and the difference between the in 1983 for end 1985 expected and by the end of 1985 realized automation degree in firms with various activities and of different sizes in the private sector.

Expected and Realized Degree of Automation in the Private Sector in 1985

	Number of Employees							
	5-10	10-20	20-50	50-100	100-200	200-500	500& more	5& more
MINING AND INDUSTRY								
expected	27	39	63	84	89	94	96	50
realized	18	32	59	80	87	94	100	43
BUILDING AND INSTALLATION								
expected	14	26	47	67	85	89	89	29
realized	10	19	41	64	83	94	95	24
TRADE								
expected	37	55	74	85	94	95	95	50
realized	28	46	67	79	87	92	99	42
COMMERCIAL SERVICES								
expected	55	59	75	87	89	91	91	61
realized	45	55	72	83	90	92	92	53
REMAINING								
expected	18	22	30	51	62	85	95	31
realized	13	18	26	41	59	73	89	25
TOTAL								
expected	35	43	57	72	80	90	95	46
realized	28	36	52	66	77	86	94	39

—expected degree of automation in 1983 for 1985

—realized degree of automation in 1985

—according to inquiries in 1983 and 1985

Degree of automation=percentage of firms with computers of at least Hfl. 10000. each and/or remployed automation personnel.

5. FLEXIBILITY, STANDARDIZATION AND CENTRALIZATION

This chapter evaluates the widespread belief in the promising opportunities for MSFs due to their perceived predispositions for flexible automation. It assesses the future impact of this type of automation for the structure of MSFs and for their position in the overall economic structure. The future of flexible automation in MSF's cannot simply be extrapolated from earlier forms of automation. The technology implies a breakthrough in its characteristic and its impact. This future cannot be derived from expert opinions either. Experts are either biased by their professional focus or their interests and they tend in practice to contradict each other.

As an alternative approach in this study a closer analysis is attempted of the specific features of the new technology in relation with those organizational characteristics, which will probably be most influenced by this technology. These characteristics are denoted by the key concepts of flexibility, standardization and centralization.

5.1. Flexibility

5.1.1. Flexibility of the technology

An important feature of electronic automation is the programmability. The electronified systems are no longer restricted to fixed functions but can perform an indefinitely large variety of tasks. The functional application is no longer fixed in the system but can be determined from outside. This gives a flexibility, which enables the same machine to switch to another production process, without drastic corrections or adaptations of the machines and thus without the corresponding time, costs and production losses. This flexibility is however not so high in practice as it is in theory. The flexibility of programmable automated systems is constrained by new limits defined by : specialized peripheral equipment, standardization of application software, the characteristics and complexity of the man-machine systems and ill compatibility of system components. In the short run too high expectations on the flexibility in general will therefore be unrealistic.

The flexibility of machine systems is bound by upper and lower limits with regard to the forms, measures, material and tolerances of the products. This limitation is not caused by the computer control but by the mechanical characteristics of motioning systems and tools and by their possibility for readaptation to new processes.

The application of standard software is another restriction to flexibility. Standard programmes require usually a high degree of formalization of standardization of production processes and organization.

An important limiting set of factors is formed by the characteristics and the complexity of the man-machine systems in which the micro-electronics and computers are incorporated. It is obvious that the synergetic effects of the coupling of systems make a high level of integration desirable. Large interdependent systems however are soon quite vulnerable, even if redundant components are built in. Moreover the costs of coupled systems and the problems of the implementation of them follow an exponential function of the level of integration. Therefore in most cases they will stay out of reach of MSFs.

Flexible automation in MSFs is, therefore, in practice still almost restricted to stand alone machines as CNC machine tools, personal computers, robots or coupled systems, consisting of a few machines and some cables or a small network. These are, however, the systems with which a higher integration should be reached. If this does not fit into an automation plan, this will meet vast incompatibility problems later, which will prevent a higher degree of integration. These limitations of production flexibilitity can be such that investments in flexible automation for MSFs can be uninteresting.

Flexibility is a desirable and an essential characteristic of computer automation. This flexibility, however, should not be overestimated. To a certain extent, for instance in small series and in very complex operations the human operator is still more flexible than any machine (system). Moreover the investments in the computer controlled production systems are still too high to compete with human operators for small series.

5.1.2 The relation of flexibility of MSFs and automation :

The widely endorsed view that, because MSFs are more

flexible than larger concerns, they are also more susceptible to innovation and for flexible automation cannot be accepted uncritically. This view suffers from fuzzy concepts and argumentation and from poor empirical support.

The concept of flexibility is complex. An abstract definition of the concept is the capability of a system to change. This change may have reference to the very system and its characteristics or to the extent, direction and speed of the change. The characteristics of a production organization, that might be susceptible to change are :

*The volume and diversity of the product range.
*Specifications of the product or service.
*Production processes and procedures.
*Nature and organization of the means of production.
*Number and quality of personnel.
*Structure of the labour organization.

It is not plausible that MSFs in all these respects would be more flexible than large organizations. In most cases the alleged flexibility means only a lead in flexibility with regard to the extent and the speed of possible changes in the structure of the organization and the distribution and assignments of tasks and of the product specifications. Flexibility in jobs and organization structure are inherently correlated by the smaller organizations, the lack of a strong hierarchy and the lower degree of specialization. However, no decisive arguments exist to assert that a small firm is more flexible in changing its production processes and machine park.

The flexibility in product specifications is not an inherent quality but a need for fast reactions to new demands of existing or potential customers. Markets cannot be manipulated by MSFs as larger concerns are to a certain extent able to. It is true that usually MSFs have more simple production processes and machine systems. They lack on the other hand the financial means to change them drastically. What will be the balance cannot be decided on logical grounds. Empirical research shows that MSFs in general are relatively fast in product innovation but not in the adoption of process innovation. But even if a flexibility of the production process can be supposed or demonstrated a remaining question is the relation with flexible automation. Admitted that the problems of changing an expensive and vast fixed automated system are larger than changing a system of stand alone conventional machines, it can be argued that firms with a natural flexibility of their production process have less reason to adopt flexible systems on behalf of the flexibility of these systems than firms with a fixed automated system. If they nevertheless adopt flexible automation the suspicion arises that they do it for other reasons ; for instance to realize a lower price per unit product. They thereby might even lose some flexibility. Therefore the causal direction of an eventual relationship can also be inverse.

Flexible automation enlarges the economies of scope and changeability without offering much to the efficiency of scale. This might lead to different motives for the adoption of flexible automation. Trends in market demands are developing towards differentiation and accelerated changes in demands. Now that the economies of scope come into reach of larger firms as the result of flexible automation they might implement flexible automation to diversify and flexibilize their product range. The production of smaller and medium series belonged traditionally to the tasks of MSFs. Competition might force MSFs to adopt this form of automation in order to retain their market shares for small and medium series. Because the advantages of the economies of scale could well be extended to smaller series by the technology of flexible automation, they thereby well might increase their productivity in producing series.

The conclusion is that the hypothesis that flexibility of MSFs by logical necessity implies a great opportunity for these firms in adopting flexible automation systems cannot be maintained in general. Empirical findings indicate that medium-sized firms, which produce larger series, tend to adopt these systems earlier than smaller firms and more specifically because of their larger flexibility.

5.2 Standardization

A number of inherent features of programmable automation tend to impose standardizing effects on the adopting organization. Automation tends by definition to run a production or information process without or with reduced human intervention. The human intervention will be reduced to the programming of the system in advance, the over-all guarding of the process and eventual trouble shooting, repairing and maintenance in the rear. Actual control and execution of these processes tend to be more integrated and autonomous. In order to prevent these more complex and autonomous processes from undesirable problems and disturbances, the process should be designed in such a way that it will be more predictable. Moreover, instruction programmes are by necessity formalized models of a real production process. For these reasons production processes need to be formalized in order to become susceptible for flexible automation.

Another reason for standardization is the strong tendency to standardization of software. Developing new programmes is a time consuming, expensive and risky activity. Existing and frequently used programmes can be delivered on short terms and at reasonable prices. They are tested and debugged extensively in operation and are thus very reliable. MSFs which normally employ no qualified and specialized programmers are in most cases thrown on this standard software for application to their production and information processes. This standard software, however, lays a strong claim to the formalization of these processes. In order to respond to this claim these firms sometimes will have to reshape, organize and streamline their operational processes drastically. However working processes with a great variety in materials, components, tools etc. for instance in unique projects and small series cannot be adapted so easily to these requirements. Many processes in MSFs appear to have a low degree of standardization. This means either a limit in the possibility to introduce flexible automation or the need for considerable changes in the organization of production. In spite of the increased flexibility based on the programmability of the computer controlled automation systems an increased formalizing and structuring of the production and labour process appears necessary for small firms in order to profit from the potentials of this technology. This implies, however, decreasing flexibility. The concrete individual weighing of the need for remaining flexible and of the advantages of flexible automation will have to determine the outcome of the option.

5.3. Centralization

5.3.1 Centralization of management and control

The control of production processes is exerted through the collection, manipulation and distribution of information. In MSFs traditionally the administration and manipulation of data has not been developed as rationally and systematically as in larger firms. These activities are partly imposed upon MSFs by government regulation or fiscal requirements, and partly undertaken as a necessity to keep insight in a growing organization. Owners and managers of MSFs are usually more interested in the primary process and tend to deal with co-porate information personally, subjectively and intuitively, They consider formalized information systems as a necessary evil and do not consciously use it as a tool of management. The need for fromal information systems and for automation of these systems only grows as the organization grows and as the management feels no longer able to oversee personally the structure and the dynamics of the organization and its relevant environment. The automation of information streams within the organization may provide the central levels with more faster and more useful information. The question is if the increased information offers enough surplus value to justify the investment. The investigation in medium-sized firms shows that this forms one of the main reasons for the implementation of computer systems and internal evaluation appears to prove that the expected surplus value indeed occurred. The new information enables central management to increase interference with the execution of work. The linking lines between management and operation can be shortened. Middle management functions, which traditionally performed vital tasks in these connecting links will either disappear or undergo radical changes. The enlarged and more immediate influence of the entrepreneur can theoretically reinforce centralization of management.

In practice the situation is not yet decided. Other things being equal, in organizations where management style is such that the management traditionally tends to take most decisions, this centralizing potential of automation can already be observed. The inherent low separation of ownership and management in MSFs combined with computer automation tends to centralize management further.

In those organizations where much information enters the organization from outside *via* the executive levels, however, the interpretation of this information will remain necessary in order to retain flexibility in the parts of the organization. Weighing the advantages of increasing central control and flexibility of the parts will result in an optimal mix between centralization and decentralization.

Finally the simplification of central management by automation of information processes promotes the possibility for scale enlarging and concentration.

5.3.2 Centralization and scale

In the automation of information processes the threshold for enlarging the scale of production systems seems to be lowered. This tendency of increasing scales is not only influenced by internal factors. Also factors in the environment and then relations with other firms are influencing the scale of the enterprises. Specifically automation of transactions and of communication play an intermediary role in this mechanism. The striving of bigger manufacturing and assembling firms towards smaller inventories and reliable just-in-time deliveries will have an effect of concentration and centralization in the various industrial sectors. Automation of communication makes it possible to provide the necessary fast and flexible exchange of information and to exert adequate control, backward, over subcontractors and—industries and, forward, over trade organizations. Necessary conditions are, however, that the firms will commit themselves to more stable relations and that the smaller firms adapt to the constraints and requirements of the total system. This is only possible if the control over the organization is centralized. The dependency relations between parties in the distribution sector will also be increased by the new telecommunication techniques. Who will profit most from these new relations is not yet clear in advance. The wholesale houses, large stores and their distribution points and the central buying organizations may exert a greater pressure on manufacturers to deliver on special orders. The retail stores might also be forced to guaranteed off-take and thus loose their independence.

In the service sector large financial and insurance organizations seems to be the promoters and early adopters in the implementation of automatic telecommunication systems. Small insurance consultants and other intermediating MSFs have the choice of being connected to these systems and thus being exposed to control and manipulation or try to do without the systems and being threatened with losing their function and market share.

6. Conclusion

In conclusion the main findings of the underlying study are:

*A relatively low diffusion and adoption of flexible automation and resulting minor and gradual changes in MSFs.

*Flexible automation is not such a big opportunity for MSFs as is often suggested.

*Flexible automation may well lead to decreased flexibility, more standardization and centralization in the organization of MSFs.

*Flexible automation may lead to greater dependence of MSFs to larger firms.

REFERENCES

(Considering the limited available space and the lingual familiarity of the majority of the potential readers only the English literature is presented).

DAVIES, S. (1979) : *The diffusion of Process Innovations*, London.

FREEMAN, C. (1971) : *The role of small firms in innovation in the United Kingdom since 1945*, London.

GOUDAS C.L. (1980) : Pande G.C., *Computers : Applications in Industry and Management*, Amsterdam.

LEWIN L. (1984): *Telecommunications, an Interdisciplinary text*, Dedham, MA 02026.

MITCHELL W.J. (1977) : *Computer-aided Architectural Design*, New York.

OECD (1982) : *Innovation in Small and Medium Firms*, Paris.

OTA (1984) : *Computerized Manufacturing Automation : Employment, Education, and the Workplace*, Washington, April.

ROBSON, M, (1985), Rothwell, R. ; *What is the role of the Small Firm in innovation in the late 1980s, European Small Business Seminar*, Chester, Oct.

ROTHWELL R. (1982), Zegveld, W. : *Innovations and the Small and Medium sized Firm*, London.

WARNECKE, H.J. ed. (1984) : *Flexible Manufacturing Systms, Proceedings of the 3rd international conference and 17th annual IPA conference*, Boeblingen, September.

Central Bureau voor de Statistiek (1986) : *Automatiseringsenquete 1985*, Voorburg.

6

New Technology and the Information Workers*

Moritz Rosenmund**

Introduction

When Marx introduced, in his 'Grundrisse...', the notion of a 'general intellect' taking over the command over the life-process of society and its transformation, he was obviously referring to the fundamental transformations brought about both by the development of the capitalist mode of production and the industrial revolution. The increasing awareness of the important contribution of scientific and technical knowledge to the development of the means of production is reflected in the marxian analysis by the conviction that science and technology not only had become productive forces in their own right but would, as incorporated in 'the (automatic) system of machinery' produce the conditions to overcome the capitalist mode of production.

The marxian account has become criticized for the Hegelian, if not teleological, tenor of the discourse.[1] And its

---

*Paper prepared for delivery at the workshop 'Social consequences of new technologies in the future work' ISA Research Committee 07—Futures Research XI World Congress of Sociology, August 18-22, 1986, New Delhi INDIA.

** Researcher, Dept. of Sociology, University of Zurich, Switzerland.

relevance as a tool for the analysis of highly-developed societies of the 20th century may be challenged by the enormous social transformations occurred during the last 130 years. Nevertheless, we believe, it has kept its interest because it represents an attempt to distinguish and at the same time to combine different levels of analysis which have to be distinguished and combined if one aims to understand transformations of work occurring in the present days.

A first essential feature resides in the analysis, at a global, macro-economic level, of the development of the 'capital fixe' (the machinery, in particular) leading, in the long run, to a challenge of the economy of time.[2] A second element relies upon the consideration of the transformation work has undergone in the transition from the pre-industrial to the industrial epoch : The 'metamorphosis' of the means of work from simple tools, mediating the transformation of materials to machinery, conceived as 'the appropriation of men's general productive capacities', points to the objectivation, i.e. fragmentation and recomposition, of work under the command of technical and organizational artifacts. This brings along, third, a reshaping of work content at the interface between men and machines, the former becoming 'conscious organs of the process of production, its observers and regulators' in the marxian discourse.

The level of global reproduction

Let us start with two trivial observations : First, the view has become widely held that the scope of enterprises introducing new information technologies differs considerably according to the particular market environments they have to deal with. The rough distinction between liberal and monopolistic market conditions may give an idea of the diversity of organizational problems to afford by means of these new technologies; as well as it may give an idea of their possible impact on the stratification, if not dualisation, characterizing the interorganizational system. Second, it cannot be overseen that the industrial sector, specialized in the production and distribution of hardware and/or software for the (micro) electronic treatment of information, has grown considerably constituting by these days a powerful and expanding sector of its own.

Hence, the introduction of new information technology takes place in a context which is differentiated in two directions; horizontally between producers and users of the technology, mediated probably by some kind of adaptors, and vertically between enterprises refferring to liberal or monopolistic markets respectively, confronted with partly divergent organization problems. We may assume that the specific conditions described by these elementary dichotomies regarding functional and vertical relationships cannot be ignored if we aim to understand the impact of the ongoing process of informatization on work in general.[3]

The first dimension has become further specified in a recent contribution by Rhona and Julian Newman (1986) who insist upon the importance of the distinction between liberal and monopolistic market conditions for the introduction and use of information technology. They argue that the crucial functions of this technology under conditions of a liberal market environment, in the tendency that of small-size or medium-size firms, resides in their search for efficiency, i.e, for an enhancement of productivity with regard to the specific goods produced by the enterprise. This scope may be attained on the basis of an economy of time and/or co-ordination and/ or control.[4] For the monopolistic sector the authors postulate an increasing need for effectivity in the multi-dimensional adaptation to complex environments. Starting with an analysis of information itself for these different forms of adaptation to market environments. Newman/Newman develop the distinction between information work, dealing with the collection and processing of information to answer specific questions, and knowledge work, connecting information with a framework that enables strategic questions to be asked.[5] In the concluding part of their article the authors draw some conclusions regarding possible effects of these distinguishing features for occupational stratification, with particular emphasis on clerical workers.

Unfortunately, however, they do not develop further the argument which, according to our own view, points to a crucial feature of technological transformations just in course. We shall pick up the topic later on again but intend to discuss first some characteristics of the horizontal interdependency between the information sector and the non-information sector of modern industry.

If we adopt the view of Manacorda (1984) we may describe the development and introduction of new technologies under capitalist conditions as an elliptic process. It is the decreasing level of productivity that motivates a modellizing of those operations which, at a given moment in history, lend themselves most to be represented in formal-abstract (scientific) and material-concrete representations (machines).

This process, aimed at an improvement of productivity of any kind of productive operations, i.e. affecting any combination of manual and intellectual activity, sets free a certain number of persons involved in that specific operations. At the same time, however, it implies a certain number of persons being engaged in accomplishing the task of rationalization, i.e. of modellizing and mechanising these very operations. This change and differentiation of the social division of labour affects both the macroscopic level of the economy as a whole and the intermediate level of single organisations. At the macrolevel it may lead to the emergence of new sectors, characterized by high rates of profit, in tendency, decreasing productivity, (e.g. the service sector in the last decades) and/or to the emergence of new substructures, generated internally by the involved organisations (e.g. technical and administrative offices). This latter development, abundantly discussed in connection with Taylor's scientific management, becomes to be explored again with reference to the new information technologies today. The 'elliptic' character of the process, according to Manacorda's hypothesis, is due to the fact that the low rate of productivity of modellizing and mechanizing aetivities themselves are to be improved by means of rationalization, as it may be observed in the case of computer industry.

1. Transformation of the organized work process at the intermediate enterprise level

If we aim to understand the impact of information technology on the organizational level, i.e. the level of integrated work processes, organized collectively, we should consider two related topics. One regards the emergence of new organizational substructures, just alluded to above, implying some professional figures dealing with the technological transformation in course. The other depends from the fact that the introduction of information technology affects the informational process, occurring in a specific organization, as a whole.

This latter fact is conceived by some authors as residing at the roots of a present 'crisis of the organization'. More precisely, as Veltz (1986) puts it : It appears that the enterprises are undergoing, to varying degrees, a general destabilization of the "norm of organization" represented in the notions of 'taylorism' and 'fordism'. As Veltz suggests, the capacities of enterprises to deal with new technological, economical and social realities, brought about with an informationalized production, prove to be very limited. There is discontinuity, according to this view, due to the fact that the hitherto coherent couple of technology and organization undergoes a fundamental transformation. The above mentioned 'crisis of the organization' esepecially manifests itself with regard to the 'taylorist' link between the norms of representation and the management of the firm/work-system (realized in a mechanized work-system focussed on the man-machine-relationship).

The "limited capacities of enterprises", we may hypothesize, resides just in the fact that the 'taylorist' norm itself becomes subject to the reshaping of the information process as a whole by the introduction of new information technologies, designed according to an informational modellizing of the work-process (as opposed to previous energetic models). If this hypothesis proves to be correct, it implies that any autonomous models of the enterprise, developed by individual workers of work groups at distance from the taylorist representation of the organization, may potentially become part of the informational modellizing of the work process.

It should be noted, however, that 'crisis of the taylorist link' does not imply its global disappearance. Rather the 'destabilization' should be conceived as the emergence of a precarious coexistence of two logics : a taylorized, traditional one, and an emerging one, based on attempts to control the process 'a distance', incorporating it in the 'logiciels' (Chanaron/Perrin, 1986). The question remains unanswered, today, whether the restabilization of the 'norm of representation' shall be characterized as a neo-taylorism or as any other form of guidance of the organized work process.

This may best be seen in the framework of the authors just mentioned. As we have seen before, if a process of rationalization is to be attained with regard to the information process there must be an adequate modellizing not only of specific work operations but, at the same time, of its implications for the global information process. Both have to be mechanized simultaneosly, i. e. integrated in the hardware and software of particular machines.

The modellizing as well as the mechanizing processes imply a thorough analysis of already existing fragments of the integral work process, tasks and operations, but also of the corresponding informational links. Hence it has to be developed at the interface between producers and users of the hardware and software to be introduced, and implies a bilateral flow of information about both the properties of the operations to be mechanized and the opportunities and restrictions of already existing technological tools.

One may reasonably assume that on both sides of this interaction a specific organization is needed for the production of necessary information and knowledge, to afford the peculiarities of a concrete organized work process.

When the techniques of scientific management were developed an enhancement of productivity by means of rationalization and mechanization could easily be acquired by an expropriation of worker's knowledge and by the administration of this knowledge within specialized, technical and administrative, structures, the 'offices' on the base of a 'mechanistic, model of the production process. (Chanaron/Perrin, 1986). With the coming of information-based technology these structures do not disappear, but their capacity to deal with the problematics as specialized agencies for the internal planning may become challenged by the very fact that new information technologies affect the global information process as well as immediately productive operations. Moreover the problems become more complex for two reasons : on the one hand, the concrete opportunities and constraints of the new machinery, usually introduced not as integral production units but as 'casual' improvements of existing working environments, and their relationships to the ensemble of conditions in a specific enterprise still remain to be explored and to be adapted in a producer-user interaction as mentioned above. The 'quality circles', discussed by Chanaron and Perrin, may be regarded as one form of acquisition of needed knowledge which cannot be provided by specialized offices alone. On the other hand the process may become further complicated under conditions of flexible specialization as opposed to standardized mass production.

With the coming of a new, i.e., informational, form of modellizing work processes, and with the development of a corresponding sector as mentioned above, the relationship between users and producers may appear reshaped : *New professional figures* are needed to deal with the problems of modellizing work processes on the one hand, and of developing and adaptation of convenient hardware and software on the other. This obviously happens on both sides of the interaction, i.e. for producers and users respectively. In-so-far the challenge resides in the modellizing and mechanising of information processes underlying the flow of production, for all contributors to its accomplishment the label '*operators of the general intellect*' may be adequate. We should add, however, that this labelling should not be reserved for a certain group of specialists. In the tendency, as the informatization (and intellectualization) of production progresses, it should be attributed to the whole range of persons engaged in the organized work process.

By assumption the introduction of new information technology leads to new forms of the social division of labour, realized by new industrial sectors, the emergence of new professional figures, or the redefinition of existing professional roles. One consequence connected with this reshaping of the occupational structure has been dealt with in a vast number of contributions to the deskilling-enskilling controversy about the future of professional qualifications. Here the issue shall not be taken up again.

Another effect should be mentioned here, however. The very activity of modellizing in the described sense may itself become fragmented according to the same logic of rationalization, particularly along the dimension between knowledge work and information work[6], but integrated around specific rationalization projects. Thus we may hypothesize a growing number of work settings with the following characteristics :

(a) Fragmentation and integration no longer occur uniquely with reference to the relatively stable organized work process of the enterprise as a whole but, at least in part, focussed on a succession of projects wherein a 'task force' with changing composition puts forward the rationalization.

(b) Specialization resulting from this kind of producing is defined collectively and within a very restricted temporal frame of reference, as opposed to a life-long perspective of individuals, be this based on professional and/or class-related features.

(c) Insofar the way of operating of thc above mentioned 'task forces' is in perfect accordance, structurally, with the rationalization targeted as the product of the 'task forces' these latter may themselves be regarded as permanently subject to the same process of modellizing and mechanizing, i.e. rationalization.[7]

The emergence of such a 'profile' of the organized work process has important concomitants, particularly at the level of industrial relations : First, the positioning of affected workers within the enterprise becomes obscured because (a) the representation norm of the organization, as structured by the principles of scientific management, loses at least a part of its strength as a frame of reference ; (b) the classical model of professional socialization, too, ceases to be an adequate description of the process of acquisition of task-related knowledge and behaviour ; (c) the fragmentation between different 'task-forces', sometimes reinforced by their decentralized localization in space, impedes an adoption of unambiguous criteria of positioning within the organization.

Second, where such situations occur the definition and articulation of scopes for bargaining with the single enterprise or a whole branch of industry becomes difficult for both individual workers and their representatives, worker's unions.[8] Correspondingly there may be a weakening of traditional bases for the development and maintenance of stable role-identities at the workplace : i.e. neither professional culture nor class-related solidarity are easily maintained within the work setting.

(2) The level of the concrete content of work

To describe the impact of information technology at the level of the concrete content of work we have to consider cultural as well as structural aspects. To summarise the relevant structural features we may follow *Gallino's* (*1985*) recapitulation of the scientific accounts. He distinguishes between principal dimensions of an 'ergonomic space', including (a) the micro-environment, (b) the interface between men and machine and (c) the fulfilment of elementary psychological needs. To this 'ergonomic space' a 'decisional space' becomes added, implying, hierarchically ordered, (a) work complexity, (b) autonomy and (c) control. As a consequence of the informatization of production a general shift is suggested towards improvements with regard to the ergonomic space, but increasing restrictions within the 'decisional space'. We may add that this quite suitable frame of reference has to be differentiated further, as soon as we consider the axis of information work *vs* knowledge work pointing just to different levels of the decisional dimension. We may see then how the introduction of new information technology may even touch activities closest to decisions about the strategies and behaviour of organizations.

If the implications regarding the distribution of power become self-evident in this case that refers to organizations, they may be more obscured, but nonetheless present, with regard to professions. As it may be seen, say, in medical work the control of the production of central goods and values by highly professionalized groups may be challenged as well by sophisticated expert-systems.

If the structural accounts of work content, changing due to the introduction of information technology, appears highly elaborated, the cultural aspects of the transformation lack sometimes a similar level of attentiveness. But, as we have argued in the previous paragraph, cultural aspects may become increasingly important with the diffusion of new information technology challenging the previously well-established 'norms of representation' and leading potentially to a proliferation of 'internal models' of work and its organization by individuals and groups. Moreover we have pointed to the strengthening of problems centered around workers' positioning and identity.

Trying a tentative distinction of cultural aspects of work content we may arrive at an ordering along the dimension of institutionalization of codifications from (a) strictly subjective codifications (e.g. work as sustaining personal identity) to (b) intersubjective codifications [e.g. work interpreted with respect to professional identity or organizational commitment] and (c) codifications connected with the societal formation as a whole [e.g. the protestant ethic].

We may hypothesize that the diffusion of information technology will challenge in a fundamental way these cultural aspects of work content. The term 'challenge' seems to be adequate, here, just because we should not expect a one-dimensional determination of codifications [or 'images of work'] by information-based technology and organization][9], but also an important contribution just of these 'images of work' to the concrete shaping of organized work.

If this hypothesis is correct it implies two important consequences, one theoretical and one methodological, may follow : fist, the impact of new information technology has to be conceived from the viewpoint both of a 'cultural lead' as well as of a 'cultural lag'. Second, efforts have to be made by social sciences to develop adequate reconstructions of the 'images of work' held by workers themselves, to enhance their capacity to deal adequately with the new technology.

Notes

1. Nevertheless the same logic becomes reproduced in the present days by several 'end of work' ideologies and other utopian or catastrophic scenarios.
2. This, however, without taking into consideration the global quality and quantity of socially necessary work.
3. This is not to say, however, that the simple four-fold classification presented here as ideal-typic, exhausts the whole range of concrete situations. Moreover, it should be emphasized that the argument is not intended to apply on an operational level in the sense, that one concrete organization always can be classified successfully in one and only one field of the table.
4. Particular importance may be attributed, today, to the coordination problem because of a global trend towards 'flexible specialization' especially by small-size and medium-size enterprises even during the stagnation of the seventies (Piore/Sabel, 1985).
5. Here, again, the distinction does not refer to enterprises as empirical entities dealing with just one aspect of the problematics.
6. This can particularly well be seen with respect to the development of expert-systems.
7. According to many authors this implies a revision by the social sciences of the current models of Industrial Relations (e.g. Tomlinson, 1984).
8. This is the view adopted by many scholars studying processes of value change, particularly regarding work values, or the developments with respect to class-consciousness.

REFERENCES

J.J. CHANARON/J. PERRIN (1986) : *Science, technologie et modes d' orgnisation du travail, Soc. du travail* 28, Nr. 1.

L. GLLINO (1985) : *Informatica, trasformazioni del lavoroe qualita del lavoro, in M. La Rosa, ed, il lavoro che cambia*, Sociologia del lavoro Nr. 23.

P.M. MANACORDA (1984) : *Lavoroe intelligenza dell'eta micro-elettronica*, Milano.

K. MARX, *Grundrisse der Kritik der politischen Oeonomie*, Frankfurt (Moskau).

R. NEWMAN/J. NEWMEN (1985) : *Information work : the new divorce* ; The British Journal of Sociology 36, Nr. 4.

M.J. PIORE/CH. F. SABEL (1985) : *The second industrial divide possibilities for prosperity*, New York.

J.D. TOMLINSON (1984) : *Economic and sociological theories of the enterprise and industrial democracy*, The British Journal of Sociology 35, Nr. 4.

P. VELTZ (1986): *Informatisation des industries manufacturieres et intellectualisation de la production*, Sociologie du travail 28, Nr. 1.

7

New Technology and Office Work

Juliet Webster*

Since microelectronic technology first became a matter of major public debate in the mid 1970s, there has been a steadily burgeining literature on all aspects of the topic. There have been explanations and accounts of the principles of micro-electronics and the silicon chip, its applications and possibilities, information and advice to the business world on topics ranging from the sorts of systems to consider for certain applications to the relative health of competing semiconductor companies, books to promote wider public understanding of the technology, and a host of glossy magazines on systems and applications. Regular articles have also featured in the major newspapers, trade and academic journals.

From the abundance of material concerned with the introduction of word processing (WP), two themes emerge which are of particular relevance for a consideration of labour process change. The first describes and isolates the benefits WPs bring, and deals with the specifics of organisational detail which may ensure a 'smooth' transition to office automation, while the second addresses the question of WP critically, regarding it as indicative of a reassertion of capitalist control over the labour process. Within the first approach it is

* Social Statistics Research Unit, The City University, London, EC1.

commonly argued that the adoption of microelectronic technology in general is crucial for the increased competitiveness and survival of the UK economy by providing new and cheaper goods and services (Department of Industry 1978; Central Policy Review Staff 1978; Advisory Council for Applied Research and Development 1980).

The benefits for office workers themselves of working with such machinery are spelt out by the WP manufacturers, who publish lavish and glossy promotional material. In the 'office of the future'—the so-called 'paperless office'—elegant young secretaries sit in air-conditioned splendour amongst pot plants, smiling into visual display units. Not only is the working environment portrayed as being cleaner, lighter and more comfortable, the work itself is held out as being more skilled and satisfying for the office worker. Overall, WP is heralded as a new "skilled profession" in which the operators themselves form part of "an integrated team, each in control of, and responsible for, their work area" (*Word Processing Now* 1978: 26). The implication is that, freed to more responsible and intellectual tasks, the office worker will become less of a machine operator and more an administrator of information, may even be able to progress through the office hierarchy into the realms of management.

Such an optimistic viewpoint seems to be the result of a conception of planned technological change as a balanced structural process and of a belief in the evolution of a new 'post-industrial' or 'information' society. This perspective, however, massively plays down the possibility of mass unemployment or increasing social and class divisions.

If by arguing in favour of the liberating, enriching effects of word processing on the labour process, it owes its inspiration to the work of Daniel Bell, by the same token the radical critical approach which has developed in response to this is demonstrably the product of Bell's own philosophical adversary, Harry Braverman. It sets out to contest the notion that jobs destroyed by the microelectronics industry will be replaced by new occupations in new industries, and that degraded and routine work will be consigned to machines, releasing human beings to 'cleaner', more creative and fulfilling work. It poses instead a view of the development and implementation of microelectronics as a reflection of the interests of capital. In the context of an ailing copitalist economy, new technology is regarded as a deliberate mechanism where by employers cut labour costs and achieve greater productivity in an attempt to solve the crisis of falling rates of profit and maintain the rate of capital accumulation. The entire rationale behind the new technology is to increase 'the trend towards the real subordination of labour to capital in the form of machinery' :

> It is within this overall context that we must view the introduction of word processors and other microelectronically based office equipment into offices—not as part of a technology which is autonomous, driving forward by its own momentum, but related to and influenccd by and crucially part of capital's strategy to continually reproduce itself (Barker and Downing 1980;85).

This perspective, then, regards the introduction of WP technology as a process which heralds the implementation of techniques more traditionally associated with the assembly line. The principal features of automated office work reflect a taylorist strategy and are those which Braverman himself regards as typical of labour processes in the monopoly capitalist era : fragmentation, degradation, deskilling, intensification and the pacing of work.

According to this argument, the bedrock of WP and the fundamental means by which typing labour processes are transformed at all levels, is its principle of task fragmentation :

> While there has been a tendency historically for the secretarial worker's job to develop as a completely manual function, many secretarial workers find they have auxiliary tasks to perform, such as filing, the responsibility for an individual's mail and making coffee. The object of word processing systems is to further fragment the secretarial job. High levels of efficiency and cost effectiveness are best achieved by stripping away all possible distractions from the actual job of typing (CSE Microelectronics Group 1980 : 48).

The secretarial worker, it is argued, is now devoted entirely to the single activity of keyboarding text into the machine and no longer sees her work through to its completed and printed state. Moreover, since documents can be stored in the machine's memory, she can theoretically work continuously at her screen without ever needing to leave her seat, to look them up in a file, for example (Barker and Downing 1980). Her confinement is thus absolute. However, not only is the office worker now portrayed as a mere 'inputter' of text, but the material she types itself apparently becomes fragmented. Standard sentences, paragraphs, letters, address lists and legal documents, for example, can be stored in the machine's memory. All that remains to the typist is to recall the material onto the screen and type in the variable information a name and address, date, amount of money, or an additional legal clause, for example.

The technical and physical characteristics of the machine are held to be crucial agents in the deskilling of typists. The capabilities of the machine supposedly not only restrict the degree of manual dexterity required of the operator by automating many of the physical functions of typing formerly performed manually, but also, and consequently, they eliminate the typist's control over her machine and her working patterns in general. We are told that the machine can produce perfect copy independently of the typist's speed, accuracy or ability to align paper, and that she needs little conceptual skill or initiative in laying out her work since the 'creative' element of typing is now performed automatically by the use of layout programmes. We are given the impression that the machine's capabilities increasingly outstrip and restrict those of the operator, who is reduced to mere button-pressing, the simplest knack, which ean be carried out unthinkingly and automatically.

The purpose, it is argued, of confining the typist to keyboarding, eliminating all 'distractions' from this activity, and creating a typing process which is simple in the extreme to perform, is to heighten the intensity of labour and thus increase output. As well, therefore, as the mind-numbing monotony and loss of skill which operator of a WP apparently suffers, she is 'under greater stress than ever before to hit keys as fast as possible, to do the most routine, tedious tasks at the greatest possible speed' (Harman 1979 : 9).

The Capitalist Labour Process and Technological change

In their attempts to counter the technological determinism which permeates the 'optimistic' approach to office automation, radical critics place great emphasis on its capitalistic character. It is seen, not as self-generating and autonomous, but as specifically developed and exploited by capital, a crucial mechanism whereby the rate of accumulation is stepped up. It often seems, though, as if the mere statement of this point is considered a sufficient challenge to the notion of the 'neutrality' of technology. I want to suggest, however, that this point is in its own way as crude and deterministic as that which it seeks to counter, and that it is, moreover equally ahistorical. The main thrust of my accusation, though, centres on the implication by radical critics that new technology in the office has as its inevitable outcome the taylorisation of secretarial work, and proceeds with a juggernaut-like logic to rationalise entire office labour processes. In this sense the technology appears just as self-propelling as it does in optimistic accounts.

My concern is to move away from a focus on the qualities of machines and the impact they have on the labour process and labourers. In my view this is a mistaken preoccupation. I want to stress, by contrast, that 'technology' can only be understood in terms of the organisation as well as the instruments of production, and that the relationship between the two has to be made explicit. It is not sufficient simply to state the connection between the development of machinery and the requirements of capital, for this is to omit the vital link of division of labour in the chain of development of the capitalist labour process, the crucial role of which is spelt out by Marx in *Capital*, Volume I.

To investigate the effects of WP, I conducted eight case studies of organisations in Bradford, West Yorkshire, representing a cross-section of local business : a small structural engineering firm, a university department, a mail order catalogue company, a firm of solicitors, a large knitting yarn manufacturer and a replacement engine parts supplier. My research indicated that, far from WP having one uniform impact on all categories of typing-related office work, several different labour processes result, from highly integrated jobs containing a variety of whole tasks to tightly fragmented ones made up of partial operations. The distinction between secretarial and typing work, and also the different degrees of the technical division of labour applied to the latter category, emerged as crucial for an understanding of change :

> The distinction between secretarial and typing work is crucial to analysis of the likely effects of WP in offices... female typists—essentially machine operators—have existed as long as the machines themselves. As specialist machine operators, typists can be managed like factory shopfloor workers; this may not be true for secretaries, who perform a range of other duties in addition to typing (Arnold *et al* 1982 : 60).

Thus, the technical division of labour in the office is the factor causing repetition; monotony and lack of social contact with authors. The following description of working in a building society shows how tasks can be simplified without recourse to automation, and that WP is not always responsible for such fragmentation :

> It was just copy typing all day. We had these cheque letters and we just typed the name and address of the person and the roll number, and that was all you did all the time. Everything you did, you just did the name of the person, and the address (. .). It was just standard letters all the time.

The rationalisation of office work and the associated fragmentation of typing tasks are by no means new phenomena brought about by the introduction of microelectronic technology. In fact, rather than organising work around fragmented operations, it seems that WP, like former office techniques, simply capitalises on processes already set in motion.

The case studies revealed an immense diversity in the extent of skill employed, and since the feature common to all these labour processes was the presence of WP, this variation could only be explained by reference to the technical division of labour in each case. Moreover, just as Braverman has been accused of constructing an over-romanticised conception of the skilled labourer as an ideal-type artisan/mechanic (Beechey 1982; Elger 1982), so too have current writers glorified the former position of some office workers, typists in particular. The ability to manipulate the typewriter has been confounded with overall control over the labour process; consequently, in idealising the nature of typing skill, it has, as Thompson has pointed out, been possible to overestimate its destruction (1983 : 97)[1]

The general picture built up in the assessment of changes in the labour process is that skill is largely based on knowledge, the unity of conception and execution, and the excercise of control by the workforce, and that the deskilling process involves these general skills being reduced to job-specific ones. However, one is forced to question the extent to which pool typists in particular ever exercised overall control based on knowledge and judgement, given their long standing situation of isolation from the source and destiny of their work. Surely their typing skills have always been 'job specific'. Many writers on WP appear to be under the illusion that such a situation is new to automated offices. In fact, it would seem that there are categories of office workers who have long been subject to the 'mental deskilling' associated with fragmented work. Indeed, the Bradford Word Processing Study Group itself admits that the typists upon which it focused (those working in the first large WP installation at the local authority) were, even before the advent of WP, 'little more than living word processors' (1977: 12).

Accompanying the deskilling trend attributed to WP by

---

1. Glenn and Feldberg, for instance, qualify their own approach in a footnote stating the importance of not romanticising traditional clerical labour (1974 : 64). Unfortunately, their actual analysis is in no way tempered by a recognition of this point,

its critics, is the drive by management towards ever-increasing working intensity and the elimination of the non-productive 'pores' in the working day. First, they argue, WP is used to impose increasing levels of work on dramatically depleted workforces. Secondly, in deskilling the office worker and removing from her the control which enables her to regulate her working patterns, it is used to pace her, eliminate porosity, and consequently heighten the effort she expands. Thirdly, an assumption is made that increased productivity has crucial implications for the intensity at which a typist works and therefore for the quality of her working day. The benefits to management imply deleterious effects for the worker and the two are written about as if they were interchangeable. In fact, a distinction has to be made between what Baldamus refers to as objective 'work realities' and subjective 'work deprivations'.

The phenomenon of work intensification is in fact far more complex than critics of the new technology suggest. The two most crucial determinants of the level of effort exercised by typists are 1) *structural*, lying in the degree to which the division of labour operates and the consequent extent of confinement of the typist to a single subdivided operation; and 2) *normative*, relying on managerial pressure on workers to exert themselves. According to this interpretation, the machinery itself plays no significant role in this process. While, as Marx himself noted, in machinofacture the worker follows the movements of the machine, which has a self-regulating moving force, this cannot be claimed of the WP operator. Her work station, when switched on, has no fixed and unvarying rhythm to which she is subordinated. In that sense she still makes use of it, rather than it making use of her, to borrow Marx's inversion. This was clearly demonstrated by the ability of all the women I interviewed to stop and start work at any time, and particularly to take brief breaks between each item or batch of work. Furthermore, despite increased productivity brought about by WP, the women were not necessarily exposed to greater subjective immiseration since WP obviated the retyping of repeated material. Claims that WP confines and paces the worker also call up images of typists as isolated individual operating units within an automated work system. Yet one of the most isolating aspects of factory work is the prohibition of conversation by the noise and speed of the machinery, and this is not a feature of operating a WP. It is perhaps more relevant to regard porosity, in individual instances, as *lacking* rather than being *lost*. Since the evidence suggests that WP was insignificant for the pace at which the operator worked, what must be borne in mind in every case is whether or not porosity was a feature of that particular job before its introduction. Only where a flow system of work prevails, whereby instead of moving about the work place to the different tools, to fetch raw materials such as stationery and to return or discuss work with bosses, the typist remain's fixed while the materials flow to her, does office work begin to assume some of the characteristics of the assembly line.

Managerial practices and worker responses

The pessimistic approach to office automation finds its conclusion in the notion that underlying the processes of fragmentation, deskilling and intensification of typing is the transference of control over the labour process into the hands of management. This follows the tradition of labour process theory, particularly that which is strongly rooted in Braverman (Zimbalist 1979 : Cooley 1980), which treats these processes as the means by which management secures dominance over the labour process. Like Braverman, pessimistic writers make no specification of the techniques that supposedly accompany WP. Instead it is implied that control is derived automatically and completely through the use of the machinery. There is a clear yet unacknowledged link with Edwards' concept of 'technical control', whereby control is 'embedded' in the technology itself (Edwards 1979 : 112).

Control in the typing pools that I studied was achieved more by the organisation of work and managerial pressure, the predilections of the individual supervisors being crucial for the subordination of the typists. This resulted in systems that were generally incomplete. Moreover, the experience of some of the typists subject to a highly detailed division of labour suggests that where Taylorism was management's preferred strategy, this operated quite independently of office automation. For example, at one building society, every second of the working day—tasks, movements, time spent on tasks, physical space even—was governed by management and here Taylorism was at its most forceful. Strict rules and regulations governed all conduct at work :

> We weren't allowed to go and make a cup of tea (. .). So you couldn't eat at your desk or drink. You couldn't smoke. You had to go into this, like a common room, like at school (. .). You could stop and talk if there was nobody watching, but you had this partition so it was a bit hard really. If you really wanted to get involved in a conversation you had to lean over your partition, so somebody would soon see. But me and this other girl who I was next to, we had this little hole in ours, so it was alright. If anybody saw you, you would get into trouble. It was very strict (Mary, Bradford Council).

Suggestions that Taylorism inevitably accompanies the move to WP suffer from a further deficiency when scrutinised. They carry an assumption that control was previously in the hands of workers rather than management, which Wood refers to as 'a kind of management by neglect' (1982 : 15), which is transformed into thorough going capitalist domination upon the automation of the office. In emphasising the role of the technology in bringing about heightened managerial control. the ability of capitalism to secure real subordination without necessarily having recourse to machines is completely overlooked. Capitalism is thus exonerated on every level.

Structures of control are more closely connected to the imperatives of the particular organisation than to the presence or absence of office automation. A perspective wherein the various capitals are rectified into a monolith obscures the particular reasons why new technology is adopted and the control structures that accompany it. The drive towards real subordination has to be empirically located within concrete contexts; this drive may be less urgent in organisations where valorisation is not the prime managerial imperative. Within the capitalist relations of production on a general level, different groups of typists are subject to varying degrees of day to day direction which can be empirically situated and explained, and which also have important implications for their *experiences* of work. Such variations are altogether ignored by the proponents of the 'pessimistic' perspective. What matters ultimately is not literal 'management' or direct supervision, which the case studies show is not always implemented, but overall control over the labour process, based on the ownership of the means of production and the ability to purchase and put labour-power to work.

Overall, my concern has been to emphasise the complexity of both control structures and the situations within which they take shape. The degree and type of control exercise by management over the office labour process is mediated by its historical use of control strategies, the position of female office workers in the labour market (including the level of wages that they can command) in particular industries and geographical areas (Bradford being renowned for its lower-than-average wage levels across all sectors but particularly in the private sector) and the state of the economy in general. The recession and the constant threat of unemployment, felt especially severely in the North of England, act as a powerful impulsion on women workers to accept their situation and affords individual capitals control over them which need not necessarily be exercised on a day-to-day level.

It seems, therefore, that it is the organisation of new technology into a *system*, based upon the technical division of labour, as opposed to the implementation of single mechanisms, that has the crucial impact on the labour process. It is *capitalism* which imparts to the production process its specific character, not the discrete mechanical components—that is, the mode rather than the means of production.

REFERENCES

Advisory Council for Applied Research and Development (ACARD), Cabinet Office (1980). *Information Technology* London, HMSO.

ARNOLD, E. et al (1982). *Microelctronics and Women's Employment in*

Britain, Science Policy Research Unit Women and Technology Series, Occasional Paper Series No 17, University of Sussex.

BALDAMUM, W. (1961). *Efficiency and Effort : an Analysis of Industrial Administration* London, Tavistock.

BARKER, J. AND DGWNING, H. (1980). *'Word processing and the Transformation of the Patriarchal Reletions* of *Control in the Office'* Capital and Class, No 10.

BEECHEY, V. (1982). *'The Sexual Division of Labour and the Labaur Process : A Critical Assessment of Braverman'* in Wood, S. (ed).

Bradford WP Study Group/NALGO Action (1997). *Word Processing : A Report on the Introduction of the Automated WP Ststems in Bradford Metropoliean District Council*, Unpublished.

BRAVERMAN, H. (1974). *Labor and Monopoly Capital*, London, Monthly Review Press.

Central Policy Review Staff (1978). *Social and Employment Implications of Microelectronics*, London, HMSO.

COOLEY, M. (1980). *Architect or Bee* ? Slough, Langley Technical Services.

CSE Microelectronics Group (1980). *Microelecironics : Capitalist Technology and the Working Class*, London, CSE Books.

Department of Industry (1978). *Microelectronics—the New Technology* London, DOI Pamphlet.

EDWARDS, R. (1979). *Contested Terraiu : The Transformation of the Workpiace in the Twentieth Century*, London, Heinemann.

ELGER, T. (1982). *'Braverman, Capital Accumulatian and Deskilling'*, in Wood, S. (ed).

GLENN, E.K. and FELDBERG, R.L. (1979). 'Proletarianizing Clerical Work : Technology and Organizational in the Office', in Zimbalist, A. (ed).

HARMAN, C. (1979). *Is a Machine After Your Jab ? New Technology and the Ttruggle for Socialism*, London, Socialist Worker's Party pamphlet.

MARX, K. (1954). *Capital*, Volume 1 London, Lawrence & Wishart.

THOMPSON, P. (1983). *The Nature of Work an Introduction to Debates on the Labour Process*, London Macmillan.

WOOD, S. (ed) (1982). *The Degradation of Work ? Skill, Deskilling and the Labour Process*, London, Hutchinson.

ZIMBALIST, A. (ed) (1979). *Case Studies on the Labor Process*, London, Monthly Review Press.

8

The Health System in 'The Third Wave'

D. Delliege*

1. KEY CONCEPTS OF 'THE THIRD WAVE'

Introduction

We intend to analyse perceptible evolution in the field of health according to the key concepts of the 'Third Wave'. A brief outline of these concepts is, therefore, necessary to start with.

According to A. Toffler, the 'Third Wave' corresponds to a series of fundamental changes in our society, the impact of which shall be as noticeable as that of the first two revolutions, which brought our ancestors from prehistory to agricultural civilization, and from there on to industrial society.[1]

These changes can better be apprehended when considered along with the characteristics of this latter type of society.

Industrial society depends on fossil fuels for energy, and implies massification and social uniformity.

Massification is obvious in mass production, mass distribution, mass education, mass media, mass leisure, and so on.

Social uniformity is to be found in the firm as well as in

* Professor at the University of Louvain, Belgium.

private life. Workers are interchangeable, and their tasks are repetitive. Products and behaviour are uniform and codified, owing to the standardization of goods, language, housing, education, working hours, lifestyles, as in the case of the nuclear family model for example.

Common principles such as standardization, synchronization, centralization, maximization and so on underlie this uniformity and diversity.

Current 'Third Wave' changes affect all the spheres of our social and economic life ; they can roughly be grouped together under a few main headings :

— General changes in society.
— Changes in production.
— Changes in consumption.
— Changes in distribution.

1.1 General Changes in Society

The essential characteristics of the evolution in progress are :

— novelty, diversity and acceleration of the change.
— regionalization.
— technical changes, particularly as far as the spreading of knowledge is concerned.
— changes in power.

1.1.1 Diversity and Regionalization

The entire scope of social, political and economic life is becoming more diversified.

The classical family model is gradually being replaced by other non-nuclear life-styles.

The range of opinions is broadening ; a multitude of small parties are appearing on the political scene, along with single-cause groups and movements such as anti-nuclear movements, supporters of abortion or advocates for foetus protection.

A variety of groups and sub-groups whose aim is to protect the particular interests of limited groups of individuals are proliferating, and continuously becoming more restricted, polarized and diversified. For instance :

— An abundance of non-profit-making organizations are emerging : Pro Vita, youth clubs, arts and crafts centers for handicapped persons, ...
— Local committees are being set up in reaction against insecurity, neighbourhood development schemes and so on.
— A profusion of recreational 'affinity groups' have sprouted, grouping together persons with a common interest in particular arts, arts and crafts, hobbies, ...
— Specialized festivals are increasingly successful, and offer a variety of themes : the cinema of the fantastic, the young cinema, the jewish cinema, the female cinema, ...
— A vast range of new sports are increasing their number of enthusiasts : Kung-fu, alkido, delta-wing flying, ultra-light motorized aircraft (ULM) flying and many others.
— Pecuilar sects such as Moon Sect, the Children of God, the Church of Scientology, to name but a few, are gaining followers.

Regional authorities are recovering specific dynamism ; they are calling for —and often obtaining – recognition of their individuality : Quebec, the Basque Country, Corsica, Flanders are all after autonomous administration, revitalized language and a feeling of pride and belonging of a nationalist type. The centralist State of the 'Second Wave' was to a great unaware of such issues of a political nature.

Simultaneously, national or multinational economies are being coupled with specificaly regional networks : Texas and Flanders make themselves known to the world at large and to potential clients under their own emblem.

Economic activity is becoming increasingly diversified inside each individual region : 'invisible' activities are developing to such an extent as to create a secondary market, parallel to that of official labour, in a society which has thus become a dual society which has thus become a dual society ; new jobs appear, various professions sub-divide into specialized branches, then split further into particular coteries and sub-specializations.

1.1.2 Spreading of Information

In the 'Third Wave', information becomes the essential good, and must be substituted for energy and manual labour.

The increased diversification phenomenon also applies to information; major dailies suffer or disappear under the pressure of a profusion of specialized journals, mini-magazines and news-letters, which fully satisfy the professional and semi-professional interest of their respective readers.

New communication media can speed up this diversification of individual information: cassettes supply specialized information which can easily be copied and circulated; cable television will give the opportunity to choose from a selection of specialized programmes, and the possibility of activity calling the emitting channel thanks to a two-way communication system. Personal, or 'home' computers, as their numbers increase, will enable firms, small businesses and private individuals to plug into fully network data banks.

1.1.3 Changes of Power

The revolutionary property of information, the new essential good, lies in the fact that it is immaterial, intangible and offers a potential set of infinite and mutually enriching uses. Access to power in the years to come will thus depend on the availability of this new knowledge and information, and the ability to use and process it. More than any other, the creation of new widespread network which can be fed by new data, is a vital means of influencing groups in general and more restricted targets in particular.

More and more information is needed with the diversification of organization, the decentralisation of poles and the growing heterogeneousness of products, values and attitudes: in turn, additional information is required to cope with coordination. A spiral of accelerated changes is thus set up with diversity generating information, and information then giving rise to subsequent changes.

1.2 Changes in Consumer Behaviour

The consumer will try and personalize his consumption.

He will favour products with additional psychological value, and will begin to play a part in the production process.

1.2.1 Consumption Personalization

The private individual now enjoys unprecedented autonomy in his decision-taking, freedom of action, and previously unknown possibilities of choosing his life-style and stating his personal opinions. As a consumer, he also wants more *customized* features in the goods and products that he buys, as well as an increase personalization of the services he uses:

—He admittedly buys mass-product goods, but endows them with special gadgets or distinctive features.

—He tends to buy specialized goods, produced in lesser numbers: specialized journals, health and diet foods, double glazing, aesthetically pleasing furniture and tiling, "soft" energy: wind power, solar power,...

—He seeks specialized services: private schooling with individual educational methods, specialized festivals, vegetarian restaurants, personal development groups, private clubs restricted to previously marginal groups (women or homosexuals for instance).

1.2.2 Psychological Perk

The consumer is also inclined to ask for a psychological perk in the goods and services he consumes: comfort of the seats in cinemas or theaters, quality of recorded music thanks to stereo systems and compact discs, comfort in the home, the psychological experience angle of holiday clubs, saunas or... zen meditation.

1.2.3 Prosumer

The consumer takes an increasingly active part in the production and distribution process: he tinkers around, buys furniture sold in kit and assembles is himself, swaps services with his neighbour, works voluntarily for local independent radio stations, and so on.

Barter and local activity re-establish themselves. Extra leisure time, soaring unemployment and the cost of official labour have further contributed to encouraging this tendency. The individual is simultaneously producer and consumer: he becomes a prosumer.

1.3 Changes in Production

1.3.1 Changes in Products

Producers adapt their activity to the new characteristics of the market; leaving the undifferentiated mass produced goods to new countries with low wage rates, they concentrate on specialized gaps and favour up-market goods: those goods that are more luxurious, more personalized, or that require deeper intellectual commitment on the consumer's part.

1.3.2 Changes in Production Relations

Decisions at the top become more complex, therefore more information is needed. Ingesting and managing this information also tends to become more difficult.

Flair and authority are no longer sufficient; a global outlook, extensive documentation and aptitudes for creativity, adaptation and negociation—all of which the Second Wave employer could do without—have become indispensable.

Confronted with better educated and more individualistic employees, the 'Third Wave' employer turns a more attentive ear to the suggestions of his personnel, encourages taking part in the firm and endeavours to create a stimulating climate. Working conditions benefit from this state of affairs : the worker's individuality is better acknowledged, thanks to personalized working hours, flextime, allotment of tasks to several persons in charge, and so on.

1.3.3 Changes in Production Techniques

Advances in technology make robotization of humdrum and repetitive tasks possible. Along with the already mentioned taking part feature, these new technologies promote:

—small units, as opposed to larger numbers of workers,
—information processing, prevalent over manipulation and muscular strength,
—creativity and enrichment of the tasks, in contrast with the breakdown of work into individual operations.

1.3.4 Changes in the Production Place

The spreading of micro-computers and telecommunications will sweep a number of duties currently performed at the place of work back to the home. This phenomenon will diminish fatigue and expenses due to commuting, and will encourage a revalorization of family life and an improvement in relationships inside the family unit and the local community.

1.4 Distribution

All the aforementioned characteristics will bear upon the forms of distribution:

—The diversity of groups and objectives require advertising campaigns to be more personalized and aimed at specific targets.

—Recent information and diffusion techniques will enable new advertising media to be used: cassettes, videos, cables,...

—The pre-eminence of information entails setting up information networks concerning available products and mentioning their charactcristics, price, delivery time, and so on. Potential users will thus have the opportunity to consult regularly up-dated information banks in direct access.

—Special atmospheres must be suggested to satisfy the consumer's need for a psychological extra; catchy packaging is no longer good enough, the cachet of the shops and quality of their staff, together with the availability of customs-out services, will prevail and should replace the rigidity of the public counter in particular.

—The surge in prosumption further implicates the consumer in the distribution process: cash dispensers, automatic petrol stations, self-service restaurants and shops, change dispensers are all examples of this already perceptible evolution.

– The revival of work at home shall also spur sales and therefore advertising at home; not only magazines, but also terminals will allow the consumer to fill out order forms, thus relieving pressure on the department stores.

2. THE 'THIRD WAVE' AND THE HEALTH SYSTEM

The key eoncepts of the third wave may be used as a reference-scheme in order to analyse the health system. Such analysis is striking : the scheme is relevant in every respect ; each aspect of TOFFLER'S general analysis may be noted in the specific field of health.

2.1 Over-all Changes in the Health System

Diversity, regionalisation, technical changes and changes of power are present in the health system as well as in society in general.

2.1.1 Diversity in the Health System

Health practice was rather standardised during the second wave : urban and rural practice, hospital and solo practice were about the sole differcnces.

The services useful for health are provided by a larger number of actors during the Third Wave : non-professionals, new professions and new types of activities.

— *Non professionals* provide useful services, such as mutual-aid groups e.g. for alcoholists or handicapped (cf. section 2.2.3); volunteers' activities : "lay-help", e.g. "hot-lines" for people in distress, information provided free of charge for specific groups (for instance, for women, for younger,...); shelters for run-aways : children away from home, battered women,...; groups helping to acquire "coping mechanisms" ; yoga, biozenergy...

—*New professions* increase in number and also the so-called 'parallel techniques' within the health sector; acrid discussions oppose their adepts to medical doctors who favor more classical therapy. A few of those techniques require long training periods and are well accepted; but others are rather marginal or even quack remedies. We will mention them all, even those which are not relevant care, because they all contribute to the 'diversity' of the therapeutical processes under review.

The most widely-known are osteopathy, chiropractice, homeopathy and acupuncture.

Some of these branches already have their sub-specialities or their 'coteries' : the unicists, multiplists, cyclicists and purists in homoeopathy, the scientific and the conservative among acupuncturists, users of visualization and of PNL (programmed neurolinguistics) in sophrology.

Other techniques are developing on the fringe of the alread existing ones ; reflexotherapy, neuraltherapy, auriculeous-therapy, acupression homoeopuncture in the wake of acupuncture ; manipulation of the cranium and vertebrae in the fashion of chiropractic.

Other approaches rely on certain methods of administration, e.g. mesotherapy or certain substances, such as : phyto-therapy, aromotherapy, insectotherapy, orotherapy, trace-element medicine, vegetarianism. Quite a few exploit the mental resources of the individual—for instance sophrology, relaxation, eutonia, hypnosis, meditation, re-birthing,—or the positive properties of certain occupations or situations, such as ergotherapy, therapy using music, laughter, bioenergy, sport, physical exercise, particularly rythmical movement, sophrological karate or heliostherapy, aquatherapy, the insulating case, natural therapy. A few even put their faith in the healing virtues of mysterious fluids : radiesthesy, mesmerism, the laying on of hands,...

Besides, *new types of activities* are developing *at the borderline* of the usual medical activities : the World Health Organization has promoted Primary Health care and Health for all in the year 2000 ; new initiatives are emerging, such as programmes of health education and people's participation : at state level, or within universities, or in local communities e.g. thanks to volunteers. Still other projects will be carried out ; for instance, health practitioners will convince 'business-minded' local and regional authorities to buy 'health projects' or 'packages of preventive measures' to promote health.

New activities and various *types of practice* may already be noted within official medicine too ; a large number of groups and subgroups now subdivide health manpower according to various professional objectives, ways of practice, philosophy of life and work... This is probably due to the very rapid increase of medical supply, at a rate exceeding by far the increase of demand ; it is probable that similar consequences will expand throughout the world together with the 'epidemy' of gluts.[2]

Organisation of work and time-sharing

Careers are diverse : a few health workers have two jobs : they are for instance, doctors as well as politicians, or 'lobbiers'. Others choose a job in a quite different sector : they become for instance, policemen or lorry drivers ! Many perform several medical activities : a private practice and occupational health or school health, or a research work or a job for insurance companies or as civil servants. In Belgium, such 'side activities' amounted to 7.5% of the work performed in 1984.[3]

Medical doctors also change the way they practise ; besides the traditional solo practice, groups favour the pooling of equipment and facilities and teams promote interdisciplinary collaboration (e.g. coordinating autonomous services or organizing centers of services.

Numerous Specialities and Approaches

The traditional somatic approach has not kept its monopoly : many practitioners are now convinced that health and health care have many psychosocial components, which should be taken into account when services are provided.

New treatments are promoted due to growing specialisation and to diversity of schools within each of them, and also to 'parallel techniques' previously mentioned.

In Belgium, twenty-eight specialities are now officially registered and about thirty combinations are recorded ; more than fifty can be listed in the world. And further developments may be expected, since new specialities might gradually become recognised in more countries, e.g. endocrinology, immunology, geriatry, ... A few specialities have already been split in subspecialities, e.g. cardiac surgery, plastic surgery,... Others are subdivided in 'school' ; the schools of psychiatry for instance number thirty (see Annexure II).[4]

Various concerns

The health professions become more heterogeneous. Subgroups strive for divergent goals. Medical trade unions for instance focus on financial objectives, with arguments based on ethical grounds. On the contrary, practitioners aiming at social and ethical objectives often use arguments based on economics to sustain their projects...What a paradox? such projects may concern : public health, collaboration with other professions and with patients' representatives, health care services meeting the needs of underprivileged groups,...

Still other groups emerge, for instance in favour of international cooperation : the Peace corps of 'Medical Doctors without borders', or the League against nuclear war. Such initiatives match with Toffler's analysis about the emergence of groups striving for specific aims : the so-called solo-aim groups. Committees preaching to palliative care for dying, those boosting the right to die with dignity, those promoting abortion and those extolling the right of the foetuses are further examples of such groups,...

2.1.2 Regionalization in the Health System

Structures of health care and decision powers in the field have been markedly reorganised in a few countries. In Belgium, in Italy and in Great Britain for instance, regional authorities have been reinforced at the expense of previous central and national powers.

Besides, regionalisation occurs at lower levels too, e.g. through local agreements ; Belgian regional hospitals for instance were required to negotiate and to agree the way for decrease the number of days reimbursed. Universities now organise further training, extra-muros : they teach off-campus, in order to meet graduates in their own regions.

2.1.3 Spreading of Information in the Health System

Practitioners and managers in the health system benefit from a large set of specific reviews and publications. Each speciality and also general practice publish their own reviews; universities and academic societies publish ; many new publications provide information focused on specific topics such as morbidity data, health consumption, professional trade-unionism, strategies of health, current political events in the health sector, psycho-social aspects of health, legal aspects of health care, social science and medicine, and so on... New techniques still broaden the scope of choice, mainly for further training : audio-visual cassettes are now available with lectures and refreshment courses.

Besides, computerised data banks now offer bibliographies classified by topic [e.g. Medlar], and information about drugs and medicines, their features and ill-effects.

Sof-ware now help identifying the proper diagnosis ; they are labelled 'diagnostic-kits'. Such computerised techniques are bound to expand with the success of micro-computers ; almost 31% of the French MDs already own one, but only 6% for professional use ; such equipment will become an actual working-tool, rather than a gadget, with the growing availability of useful software.

Available statistical data of well-being or factors of well-being, and computed indicators of the kind will be integrated in systems of information, useful for decision makers when they need information about time trends or regional distribution.

2.1.4 Changing Powers in the Health System

The distribution of power within the medical profession might change markedly in a near future. Local MDs admired for their ability to diagnose promptly and to treat adequately might well lose their fame. Future leaders might be the few 'super-brains' who will be in a position to build up, to word and to up-date continuously the softwares of decision trees about symptoms, tests and procedures conducive to diagnosis, and proposed treatments ; the crowds of other medical doctors will only use such software, in a rather standardised and automatic way.

The terms of practice for the medical professions might evolve thanks to the information techniques of the 'Third Wave'. Medicine was still recently practised with methods sharing features of the first wave : at a local level, on a small scale, and with output often due to the hazards of nature ; now terms of practice might become more standardized, thus sharing a feature of the second wave. Nevertheless, diversity will be the rule again, thanks to various data banks, diverse conception of care, and the usual competition between universities and countries aiming at developing their own tools. Consequently, the third wave practitioner might be compelled to manage a lot of information and to make his own choices. How will medical doctors attempt to reach self-accomplishment in such a new environment is a matter of debate, that we will discuss in the conclusion of this paper.

2.2 Changes of the Patient's Behaviour

The patient is a consumer of health care and, just as other consumers, he wishes more diversity, he requires attention for his psychic needs and he becomes more involved in the process of care, acting namely as a 'prosumer'.

2.2.1 Diversity of Health Care

Specialists' services are used more frequently ; at an early stage, increase occurs together with a growth of GPs' services ; at a later stage, a threshold is reached and the increase of specialists' services occurs at the expense of GPs' services (see graphs in Annexure II)[5]. In Belgium for instance, the utilisation of GPs' care is increasing only slowly since '81-'82, even becoming negative occasionally, whereas specialised medicine still grows at a rapid rate (2.4% per year, conducing to a doubling within thirty years, would the increase keeps its present rate).

The proportion of specialists' services is thus increasing (27 % in Belgium). The choice is often made by the patient himself rather than by referral : in 1975, the initiative was taken by the patient in one case out of two, and even in three cases out of four, for pediatricians and gynaecologists.[6]

Another indicator of the patients' increasing taste for diversity of care is the infatuation for non-official types of treatment.

Who are those patients choosing non classical types of care ? We have interviewed medical doctor working in these fields : they did not think that their own patients differ-markedly from 'classical' patients. Actual differences may exist though and MDs did mention a few peculiarities, for instance :

— an overrepresentation of a few groups ; for acupuncture for instance, they mentioned :

— functional diseases, anxiety. nervousness, psycho-somatic troubles,

— pains : neuralgia, rheumatoid diseases, muscular aches and post-surgery pain,

— allergy,

— chronic diseases, as opposed to acute ones.

— an overrepresentation of young patients belonging to a rather high socio-professional class (one doctor stated : professionals, sportsmen and artists,...) and underrepresentation of children, elderly, and low-income patients.
— a type of patients with higher expectations and a positive and collaborating attitude towards care.

Such statements sustain the hypothesis that psychosocial factors influence the utilization of such care ; however, data are not large enough to allow final conclusions.

The patients' satisfaction seems to be high; for specific reported diseases, patients thought that the efficacy was higher than within official medicine[7] ; here again, good controlled data is lacking, because patients answering to the survey were not randomly selected.

Why do patients choose such parallel treatments ? Simple curiosity or deliberate choice may occur but they are not the single reasons : 'very often' (the exact percentage has not been assessed), patients are disappointed by official treatment, which is thought to have had no result or which could not meet their problems, nor answer to their expectations. Both patients and medical doctors have mentioned this two-step process. Further hypothesis may be proposed, to explain the present infatuation for alternative types of practice. Values, techniques, working conditions and ways of life, change rapidly nowadays, and people are compelled to adapt consequently ; the induced strees may provoke physiological and psychological reactions, and extra-utilization of medical care. Solving existential problems is now a matter for the health system, because such solution is socially acceptable. People may look for the salvation of the medical knowledge or its substitutes, because the hardship cannot be sublimated through a relationship with a transcendent being, and because no religious mediator is considered useful. An existential and metaphysical demand is thus transferred from the Church to the medical practice or to the healer.

2.2.2 The Psychological Perk Required by the Patient

The patients gradually require a psychological perk, as can be noticed by a few indicators, such as the type of consultation required, or the content of the demands.

When the patient reaches a certain threshold of well-being, he rejects the queues and the waiting lists and he prefers to go to a private practice.

Acute disease is not the sole reason for encounter : many patients may now look for a psychological well-being, or even for a piece of happiness, be it an illusion,

Aches and psychological problems were reported to be overrepresented among patients looking for 'parallel treatments' ; this is probably due to the alleged usefulness of such treatments for the mental well-being and the ease of the pain (e.g. with relaxation, sophrology) ; other techniques, such as bioenergy act as exutories of repressed feelings or energies ; a few ritual ceremonies of ancient civilisations fulfilled the same purpose.

Quality of life is also required within official medicine :

— Analgetics, tranquillisers, euphoria-inducers are massively used, to a point where critics and fears arise.

— Sleep-inducers are used as long term treatments rather than as occasional help.

— 'Non-violent' birth aims at decreasing the physical and emotional shock for the bady during his/her transfer from a liquid environment to open air.

— Palliative care for dying promote the care of the pain, emotional support and a preventive work with the families,

— The controversial claim for abortion, is also a piece of this search for a certain 'quality of life', for people claiming the right to their own well-being.

2.2.3 The Patient as a Prosumer

Self-medication is not new... It is a kind of 'spontaneous prosumerism' (see section 2.4.4 for a new type of 'induced prosumerism').

Spontaneous participation increases, namely through the trend of 'self-care' and mutual-aid and through community participation.

'Self-care' aims at inducing the patient to take more responsibility, thanks to techniques, such as relaxation, self-injection of insuline, vegetarianism,... The trend might still move up, thanks to the success of 'ecologist' ideals ; will garden-phytotherapy expand and will there be more home-made infusion syrups and shampoos ?

Mutual-aid groups gather people suffering from similar diseases : weight watchers, alcoholists, disabled, patients having undergone laryngtomy, mammectomy, renal dialysis,... Members of such leagues share specific information and provide mutual support. Such information and support have a higher credibility when it comes from 'peers' having to bear similar problems.

Finally, there are patients' committees and community participation; in a few countries, such as Canada and Great Britain, such committees are numerous. Their impact has varied markedly, according to the leaders' personality. A few of them were actually politically oriented, thus inducing negative reactions. Others have successfully created an atmosphere of trust and collaboration, allowing thus the setting up specific activities within local communities. Trends are not obvious; when participation and dialogue match with the traditions, it might successfully help to the promotion of health. In societies, where strong management is the rule, managers might oppose outsiders ; in Canada for instance, managers have attempted to gradually decrease the possible links between professionals and users lowering thus people's participation.

2.3 Changes in the Production of Health Services

Changes are noticeable in the type of available care, in the social relations, in the techniques of production and in the places of production ; such changes match nicely with A. Toffler's forecasts.

2.3.1 Changes of goods and services in the health system

New goods and services become available : non-professionals contribute, non-official medicine is expanding and much diversity has been introduced in official treatments (see section 2.1.1 for further comments).

Let us examine here a few specific changes of goods and services caused by technical evolution of production.

In the past, possibilities for diagnosis were limited. The types of care were less numerous, more invasive and non-selective ('blind' prescription contributing to further diagnosis).

The range of diagnosis techniques now becomes wider and new technologies are often less invasive (echography, CAT-scanner). Consequently, there has been a large increase of 'test for diagnosis' : radiology, lab-tests, nuclear tests ; a two-fold increase between 1974 and 1979 in Belgium for instance, was markedly more rapid than for other types of care.[8]

The increase has been especially rapid when computer techniques replaced manual labour intensive and small-scale work and provided much wealthier information than in the past. This is the case for nuclear medicine : the use of digital processing enabled the calculation of dynamic parameters and the computerised analysis of pictures ; consequently, the 'in vitro' nuclear tests in Belgium reached in 1982, eighteen times their level of 1974.[9]

Consequently, the cost of insurance schemes is increasing and attempts to control such costs are noticeable in many countries.

The range of medicines has also become larger ; more

complex combinations of substances allow the treatments to be more specific.

A promising technique is now available for the pharmaceutical industry : coating the active substance in a biological envelope, which would search the precise cells supposed to be treated ; further in the future, specific envelops might be ordered for individual patients, after due specification of their own biological features. Here again, care made to measure would replace standardized treatment.

The industry of medical devices might also expand, thanks to progresses of new techniques :

— new devices will allow self-monitoring of treatment (cf. 2.4.4.) ;
— computers and softwares will facilitate the search of precise diagnosis and the choice of relevant treatments ;
— more prothesis will palliate the deficiencies of ear and eye, and be used as substitutes for lost members or ill-functioning internal organs, such as the heart or the kidney,...

Large prospects also concern the possible uses of digital processing in health care ; in Belgium, a group of experts, nominated by the Ministry of Scientific Policy has addressed to the Government a report about such prospects and their possible socio-economic consequences.[10]

2.3.2 Changes of Relations in the Process of Producing Health Care

Changes are noticeable within the hospital setting, and also between professions as well as between healers and patients (healers meet the patient-consumers while they are 'producing' or 'providing' services).

—Participation within the Hospital

Medical doctors have large responsibilities in hospital ; their role at the managerial level vary ; in Belgium, they have to contribute to management : this compulsory role has been enforced by law in 1986.[11]

Other health professions are not involved in the decision process despite a proposals made by the Minister of Health twelve years earlier.[12] This project had also proposed to include people's representatives (e.g. general practitioners, delegates of Health Insurance Funds...) in "Watch-bodies" similar to those advocated in industrial relations at the European level[13]; such project was never adopted.

—Collaboration between Professions

Nurses' status is changing too; in Belgium for instance, regulations enact their specific role and list the nursing services for which they are qualified[14]. Less formally, staff meetings facilitate the dialogue as well as team work within hospitals, medical centres or groups practising coordinated care.

A few graduated nurses become 'entrepreneurs'. They organise personally their working conditions, in close connection with other health services.

—The Patient : a Relevant Partner

Also the patient is increasingly involved in the process of care. Many papers and books had previously criticised hospital care, using striking expressions such as : the patient as a number, as an object, as a plain body examined by students and teachers disregarding the presence of a human being; they had described the obscure jargon used at the bedside and the 'mute' patient who was hardly conveyed to take part in the discussions about his case.

Reacting to depersonalized care, new initiatives aim at promoting patients' active participation in the therapeutic process. Balint has brilliantly put forward the importance of the therapeutic relationship in the process of care; groups of medical doctors and other care providers are using his advice around the world. Initiatives also aim at providing in-patients with more rights regarding information and dialogue ; a 'patient's charter' has put such rights on paper and it became compulsory for public hospitals in France to hand it over to admitted patients. Besides, many hospitals now enquire about the patients' opinion and ask them to assess their hospital stay. Committees of patients have been set up and they collaborate and discuss with the team providing care. Last but not least, health education is progressing too; aiming at promoting patients' autonomy and responsibility, for instance through healthy life-styles, preventive behaviour, self-care and self-regulation of care; such initiatives increase personal involvement and gradually promote levelling relationships between health practitioners and patients.

2.3.3 Changes of Techniques and Organisation Process

Diagnosis techniques evolve as mentioned earlier (section 2.3.1).

The way of providing care has also changed; new working conditions influence the 'production process'.

Besides the traditional solo practice there are group practices or teams of practitioners prone to collaborate with colleagues or even with other professionals (for example, through coordinated care). Organising duties during nights and week-ends, combining professional responsibilities and personal or family activities (for instance, through a four-day week) are facilitated by the growing number of health graduates.

The production process has also evolved within hospitals. The patients' rooms become more comfortable, as far as previous large wards with twelve or twenty patients are replaced by rooms of two to four beds. The average size of institutions may have increased (as in Belgium for instance) but there is a limit ; an excessive size induces drawbacks that may be palliated thanks to internal segmentation : management is then shared among autonomous departments responsible for their own budget.

Many industrial wards have abandoned line-work in favour of more complex and significant tasks ; the nursing profession too is enjoying the so-called integrated care, in contrast with the specialised and very specific tasks previously assigned to each individual nurse. Could such type of organizing work help decrease the 'burned out' syndrome which has induced many health workers to quit job and to look for alternative types of care ?

Production techniques even change for the MD's training, which usually was a mass-training, provided in huge auditoriums ; nowadays, a few medical faculties, e.g. the University of Maastricht in the Netherlands only work with accompanied self-learning : the students receive a working programme and a list of problems to solve, either in the libraries or with the support of teachers who must be available for answering their questions but who do not give traditional courses. Computerised types of teaching might help promoting such initiatives. Students being able to follow most of their training far from their teacher, and eventually far from bedside ! How will the therapeutic relationship be taught in such a technical context ? Standardized psychosocial anamnesis may well be drafted, but the interpersonal relationship, which has healing virtues in the therapeutic process, remains to be learned on the spot, with the help of TV scenarios.

2.3.4 Changes of the place of practice

Decisions are being decentralised : from central to regional (see section 2.1.2), from regional to local and from the practice to the patients' homes.

From regional to local

The importance of Primary Care has been put forward in the famous Alma Ata Statement, together with a health responsibility at local level and the participation of population for health promotion. Such trend favours small-scale dispensaries and numerous local centres and volunteers' work, rather than larger institutions centrally located ; the concept of 'care' has also been broadened ; components of the overall development which are essential for people's health (such as food and potable water are now included).

—From the MD's practice to care at the patients' home

In this field the model is not confirmed : the share of

care provided at the patients' home has been slightly decreasing, at least in Belgium (from 56% to 53% of the GP's contacts between 1970 and 1982).

Home care might increase in the future though, thanks to a policy of keeping patients at home, rather than in hospitals. Such type of organisation has been proven to be efficient when it is feasible ; for instance, the regional hospital rate may decrease and the cost of home care be lower ; in a community under review, savings could be assessed to reach 20 $ per inhabitant in one year[15]. The trend will also be facilitated by new techniques such as :

—centrally located watching boards, where people can easily address an alarm signal.

—remote control of vital functions by computers, such as for cardiograms.

—heavy care managed at home, thanks to movable devices, such as kidney-dialysis (provided hygiene and equipment are sufficient).

Other types of services at the patients' home are also expanding, for instance "hot lines", providing support for people in distress and aiming at preventing ill-health consequences and suicides.

In all such cases, new technologies facilitate care in close connection with the usual social surrounding of the patient rather than technical institutional care.

2.4 Changes in the Distribution of Health Care

The evolution mentioned above will undoubtedly influence the distribution of care, for instance : the segmentation of the customers in the pharmaceutical industry, the use of information, the psychological perk required by consumers, the increasing 'prosumerism' and home consumption.

2.4.1 The Diversity of Customers

As mentioned earlier, patients' expectations are diverse and so are the types of care being provided and the groups and schools within the medical body (see section 2.1.1).

The various subcategories of medical doctors now form as many separate targets for the pharmaceutical industry. Will the latter adapt its 'message' to each group ? Will it focus different types of publicity according to features of each group ? Will efficacy be used as the main argument in very technical wards, whereas a more 'natural' and 'relational' context would be favoured for psy-oriented healers ? Will health products also obtain a personal touch and be produced 'to measure', according to features of the consultation and to patient's biological specificity (such as is already the case for specific anti-allergy vaccines and for homeopathic treatments) ?

2.4.2 The Use of Information in the Health System

Now that micro-computers are sold all over, there will be more data banks straightly available to the population.

First, computerised data banks will provide general information about health services (institutions, professionals) and about non-professional services useful for health (namely mutual-aid groups, etc.) : addresses, prices, age and sex, availability in time, waiting lists, features about the type of services provided, might thus become of immediate visibility. Just as the 'Michelin' rosters have labelled hotels and restaurants, there will be rosters about the health system, allotting stars and levels of comfort for each producer ! Competition will then be based on 'psychological aspects' of the treatment rather than on the sole efficacy ; the danger is that actual efficacy would probably be neglected in such lists !

The pharmaceutical sector has already set up data banks about substances and products, including names, packing and prices, therapeutical effects, side-effects and possible dangers. Straight access by terminal will become the usual connection, rather than the telephone. Later on, the patients will also be connected ; they will first obtain just "general" information, such as prices or side-effects,... In the future, information about therapy will also be disclosed : it would be an illusion to believe that professionals will keep a monopoly of access to information about therapy. Here again, new technology might fill part of the gap between professional knowledge and lay people, at least for those lay people able to afford and to use them.

2.4.3 The Psychological Perk

People's expectations in the field of 'total' well-being will be better met : the number of services dealing with psychological and social aspects of care is increasing, there will be more 'personalised' techniques and drugs with a more personal touch, places to provide care will be more decentralised, .. All those trends share one feature in common : they meet some of the patients' new expectations.

The success of 'non-classical' treatments seem to share the same purposes ; medical doctors who practise them explain the new infatuation, first by 'the therapeutic efficacy' (this is a normal rationale) and second, by their patients' specific needs to be considered 'partners' and to be treated as total human beings.

In the field of distribution, we can still mention :

—The will to favour accessibility of care : decentralised medical centres were built, cheaper than private practice and smaller than out-patient departments in hospitals.

—Waiting rooms are becoming more comfortable, with low noise, nice arm-chairs and sweet music.

—The patient's anamnesis is more often extended to psychosocial aspects of health and health care.

—Health education is evolving : from purely technical information to an attempt of 'conditioning' people, using arguments about motivations, values and feelings, just as marketing techniques advise to advertise for commercial products.

2.4.4 Prosumerism in the Distribution

Prosumerism is spontaneous (see section 2.2.3), as well as

induced, now that the producers favour co-distribution, and coproduction. Such trend is of overall applicability in the services : for self-services restaurants, for cash-dispensers, for automatic petrol stations. Why would the health sector avoid such a trend ?

The patient might act as co-distributor, for instance, with automatic drug-dispensers. Common drugs would thus become available day and night, just as gasoline, bank money and bread.

As co-producer, the patient will be invited to cooperate to diagnosis, monitoring and care, as brilliantly forecast by the French President's adviser J. Attali[16] ; industries dealing with care will open up new channels for trade : new products will be found and made available straight to lay-people. New types of behaviour will be promoted, such as 'self-monitoring' and 'hypermonitoring'; patients will be induced to buy more products, more gadgets and sophisticated devices, to check on their own health. This is already the case for a few diagnosis, such as tests of pregnancy, the reliability of which is increasing, together with their precision and precocity. Appliances to measure blood pressure are already sold straight to consumers; will they become as numerous as the thermometers ? 'Bio-feedback' gadgets allowing each one to check on his own muscular stress are already available and distributed for advertising purposes. Will there be appliances facilitating the regulation of diabetes ? Other fittings will be proposed in the future : for measuring, the pulse, for analysing blood and urines ; results will be displayed on micro-screens, together with their significance, since they will be automatically compared to reference curbs, according to age and sex... Will there also be fittings aimed at monitoring periods of fertility for women, or curbs of reactivity during the cycle ? Will E.C.G. with digital processing of the data be sold straight to patients, allowing self-monitoring rather than remote control ? Will diagnostic kits be sold to lay people ? Why not ? Everyone knows the success of Dr. Spock's book on children's care. Similar information will be sold on magnatic tape. Why would companies hesitate to expand such information to adults' symptoms, if profit seems ahead ? Even if such data banks are first restricted to professionals, who are responsible for care, the monopoly will not be held : tapes will first be copied illegally and later on they will be officially sold.

Patients will also be associated to care 'production'. This is already the case for diabetics who inject insuline, or for allergic patients who give themselves their own immunisations, or for people attending 'psy' groups or practising co-counselling, every member being a patient as well as an adviser for his/her partner. Will all those techniques spread on a larger scale and for more treatments in the future ?

2.4.5 Coming Back Home

The greater favour of home care requires that more non-professionals become involved in the treatment : on the one hand, the family, which has to accept the extra charge of caring for patients at home, on the other hand, volunteer workers who have to organise and to run services which are not funded by public grants, for instance errands or meals-on-wheels or home-help to clean-up the home,...

Conclusion

The evolution described by A. Toffler covers the biggest steps of human history, in its search to monitor factors of its well-being : first mastering nature (using fire, agriculture, tools and the energy of animals), then increasing average wealth through mass-production and more attention to the body (mastering reproduction process, prolonging life, trying to improve the aging process,...) ; the new technologies to-day provide new tools to humanity ; will they be a help to master other factors of well-being, namely social relations and mental well-being ? It is probable ; computers might become strong tools to obtain easily, quickly and permanently opinion polls on various matters, it will be a powerful way to communicate between the decision centres and the basic people of any group. New channels of decision will have to be invented, so that minorities keep a say...

Such evolution has a striking parallel in the way individual human beings wish their needs to be met at work, according to Maslow's theory of needs. First they aim at satisfying their basic physiological needs : to be fed, sheltered, and clothed ; this corresponds to Toffler's first wave for humanity. Then, they aim at obtaining security : security at work, to avoid industrial injuries and security of employment, to avoid the distress of being out-of-work ; this corresponds to the period of industrial regulations and social security and matches with Toffler's second wave. According to Maslow's theory of needs, people then aim at harmonious social contacts and esteem ; at the top of the pyramid, they then look for self-accomplishment. The ideal self-accomplishment has also markedly changed over time : first within and thanks to the basic group, each role being strictly codified from birth onwards then above the group, the individual attempting to perform "better" than the average peer ; mobility beomces the rule ; large satisfaction can be attained for those who succeed, but frustrations are as profound for those who do not. Now, will the third wave induce a dual society ? Part of such society would remain poor and would look for values of conviviality rather than individual achievement, the other part would master the information process, would contribute to its development and would thus have access to the new types of wealth. In the final search for self-accomplishment, two possible ways are possible ; the noticeable trend nowadays is a search for comfort and a taste for hedonism and individual pleasure ; when we look at history of societies, the search for pleasure and hedonism has usually led to decline and fall of empires ; but we can also notice further periods of spiritual and altruistic values. If we transpose to the evolution of human being, we can also forecast a shift from hedonistic value as ideal of self-accomplishment as it stands nowadays, to spiritual values, being considered the supreme stage of mental well-being and accomplishment.

Coming back to the evolution of the health system, similar stages may be noticed :

- First medicine is an art rather than a science ; the diagnosis is based on straight body examination ; remedies are small-scale mixtures, made out of local herbs ; causes of diseases are often ascribed to external causes and attempts are made to calm the 'spirits' and to obtain cure thanks to their intervention; it is the 'First Wave'.

— Then care becomes more technical and scientific : machines are used for tests and diagnosis ; health workers gather in large hospitals, where patients are treated massively, and drugs are prepared on a massive scale by the pharmaceutical industry. One cares for the body and its organs. It is the 'Second Wave'.

— Information and participation become more important. The patient is associated to his own treatment. Data is stored and used thanks to digital processing ; the healer is helped in his search of diagnosis and prescription by softwares conceived by experts ; this part of his 'art' being now accomplished far from bedside, the medical doctor will look for self-accomplishment in his function by caring for more subtle aspects of well-being ; he will work as a plain medaitor for applying standardized scientific information as traditional healer for aspects of well-being which are not easily codified in software ; he will become the confidant of human beings for mental well-being. He will look for self-accomplishment in this difficult role which is to be learned : to help other on the way to their own self-accomplishment.

REFERENCES

TOFFLER, A., — *Le Choc du Futur*, Paris, Denoel, 1971, 539 pp.
— *La Troisieme Vague*. Paris, Denoel, 1980, 663 pp.
— *Les cartes du futur*. Paris, Denoel, 1933, 267 pp.

D. DELIEGE (1986) *Imbalances in Health Manpower : Recent Evolution and prospects tn Europe*, Paper presented at the XIVth CIOM — WHO Conference, Acapulco, 7-12 September pp. 50.

X LEROY (1985) *Demographic: activite et revenus des medicins belges in 1984*, Document, 26 tableaux.

American Psychiatric Association (1980) a psychiatric glossary, 5th edition, Ed. by WERNER, A., e.a., IX + 142 p. (table reproduced : p. 83)

X. LEROY (1981) *L'acces aux soins medicaux*, Tome I, Bruxelles, services de la Programmation de la Politique Scientifique, pp. 229-231.

Eerstelijngezondheidszorg, (1978) Programmation de la Politique Scientifique, Bruxelles, Rapport IC, pp. 88 et 306.

La parole est aux patients — Medicine classique et therapies paralleles, Test-Achat nr 251, dec. pp. 4-14.

X. LEROY (1979) *L, acces aux spins medicaux*, Tome III Donnees regionales d'offre eu de consommation en p. 67.

A. BOUCKAERT et X. LEROY (1984, *La diffusion d'une technique de pointe : le cas de la medicine nucleaire en Belgique*, Bruxelles, pp. 1 et 183.

D. DELIEGE — W. SANSEN (1981), *Les consequences socio-economiques du development de la micro electronique en Belgique*, rapport stencile, Conseil National de la Politique Scientifique, Bruxelles, 119 pp.

Arrete no 407, 18.4.1986, (Moniteur Belge 6,5.1986).

D. DELIEGE et X. (1978) LEROY, *Humanisons les hopitaux*, UCL, Geneve, Institute Sandoz d'etudes en matiere de sante et d'economie sociale, Paris, Maloine, pp. 220-223.

European Economic Community, Commission, *participation des travailleurs et structure des societes*, Bulletin des Communautes europeennes, Bruxelles, supplement 8/75.

A. R. du 11.3.1985 fixant la liste de prestations techniques de soins infirmiers et d'actes pouvant etre confies par un medecin aux praticiens de l'art infirmier, les modalites de leur execution ainsi que les conditions de qualifications requises, Moniteur belge du 22.3.1985.

X. LEROY et L. GUNNING, *La coordination des soins de sante Primaires, Une alternative a l'hospitalisation*, Rapport au Ministre de la Communaute Francaise, Bruxelles, 1982, 86 pp. + 5 pp. + 5 pp. annexes.

ATTALI, J., *L'ordre cannibale*, Paris, Grasset, 1979.

LIST OF ANNEXURES

Annexure 1 : List of schools of psychiatry.

Annexure 2 : Amount of GPs' services according to the level of specialists' services.

Annexure I

LIST OF SCHOOLS OF PSYCHIATRY

(*i*) **Reconstructive**

A. Psychoanalysis—Sigmund Freud

B. Neo-Freudian, modifications of psychoanalysis

1. Active analytic techniques—Sandor Ferenczi Wilhelm Stekel, the Chicago school (especially Franz Alexander and Thomas French)
2. Analytic play therapy—Anna Freud, Malanio Kloin
3. Analytical psychology—Carl Jung
4. Character analysis, orgone therapy—Wilhelm Roich
5. Cognitive—Jean Plaget
6. Developmental—Erik Erikson
7. Ego psychology—Paul Federn, Eduardo Welss, Hoinz Hartmann, Ernst Kris, Rudolph Loewenstoin
8. Existential analysis—Ludwig Binswanger
9. Holistic analysis—Karen Horney
10. Individual psychology—Alfred Adler
11. Transactional analysis—Eric Berne
12. Washington cultural school—Harry Stack Sullivan, Erich Fromm, Clara Thompson
13. Will therapy—Otto Rank

C. Group Approaches

1. Orthodox psychoanalytic—S. R. Slavson
2. Psychodrama—Jacob L. Moreno
3. Psychoanalysis in groups—Alexander Wolf

Source : A psychiatric Glossary, ed. by the American psychiatric Association, fifth edition, 1980, p. 83.

4. Valence systems—Walter Blon

(*ii*) **Reeducative and Supportive, Individual and Group**

1. Client-centered (non-directive) —Carl Rogers
2. Conditioning, behaviour therapy, behaviour modification
   a. aversion therapy—N. V. Kantorovich
   b. behaviourism—John B. Walson
   c. classical conditioning—Ivan Pavlov
   d. operant conditioning—Burrhus F. Skinner
   e. sexual counselling—William Masters, Virginia Johnson
   f. systematic desensitization—Joseph Wolpe
3. Cognitive behaviour—Aaron Beck
4. Family therapy—Nathan Ackerman
5. Gestalt—Wolfgang Kohler, Kurt Lewin Fritz Paris
6. Logotherapy—Victor Frankl
7. Psychobiology (distributive analysis and synthesis)—Adoll Meyer
8. Zen (satori)—Alan Watts

Annexure II

AMOUNT OF GPs' SERVICES ACCORDING TO THE LEVEL OF SPECIALISTS' SEVICES

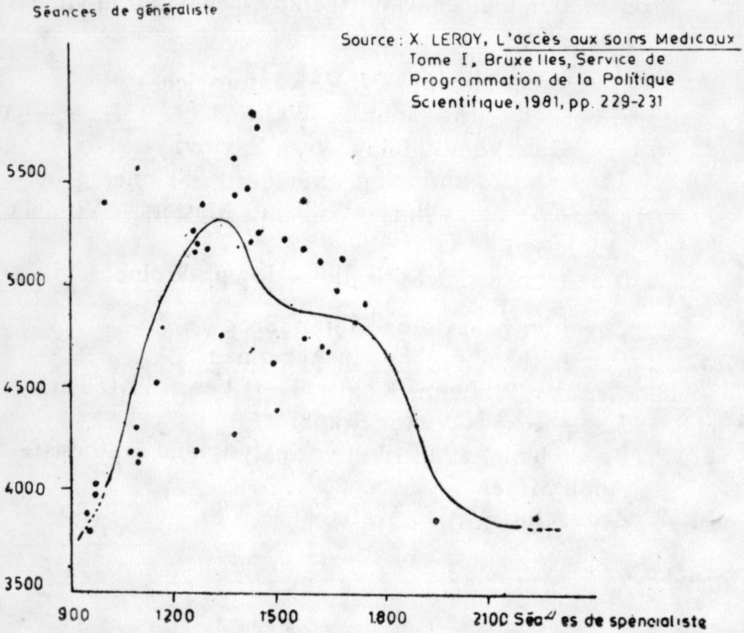

Average Utilisation Rates of GPs' Services in Six Groups of Districts, Ranged According to Utilisation Rates of Specialists' Services (standardised rates per 1000 population, Belgium, 1974)

Groups	Specialists' Consultant per 1000 Population	Number of districts	Population X10s	Medical densities of GPs +Special (amb.+hosp)	of GPs Ambul.	of Special (amb.+hosp.)	GPs' calls (average of standardised rates)
1	< 1 125	10	1.1	8.9	4.8	3.9	4 341
2	1 125—1 289	9	0.9	10.4	6.5	3.7	5 114
3	1 290—1 455	9	2.1	11.3	5.8	5.3	5 295
4	1 456—1 622	8	2.4	13.8	6.0	7.4	4 928
5	1 623—1 788	5	1.8	16.7	7.0	9.0	4 910
6	< 1 788	2	1.6	20.9	6.4	13.6	3 862

9

*Trends of Development in the Highly Industrialized Societies**

Anna Coen**

1. THE CRISIS OF ECONOMIC PLANNING AND BUDGETING

A recent long period of 'zero-growth' of the GNP made budgeting critical, while the public revenues cannot increase as much as the entitlements and the expectations. Therefore, a zero-growth phase showed that planning cannot mean just economic planning, and especially budgeting. And the ongoing crisis of the welfare state is the open facet of the crisis of budgeting in those highly industrialized countries where, during the growth (years 1948-1968) planning was mainly 'economic' planning and, in some cases, was just budgeting.

---

*Paper discussed at the meeting on "the social consequences of the new technologies" organized by the International Sociological Association's Research Committees 7 (Futures Research) and 9 (Social Practice and Social Transformation); University of Zurich, August-September, 1984.

**Dr. Coen is researcher for ISPE (Istituto Studi Programmazione Economica, Roma corso Vittorio Emanuele 284); she is also Vice-President of the Committee 7 "Future Research" of the International Sociological Association; and is member of the World Future Studies Federation.

Besides the material, financial limits for the budgeting and the welfare state, there are basic, intrinsic limits to the planning by budgeting. As much as a society becomes rich and complex, as much the needs differentiate and change from one group to another, among persons and also in one person in different periods of life. And the increasing communication, characterizing the developed countries, multiply the models among which a person will choose his or her way of life or life-style.[1]

Therefore, economic planning by budgeting cannot reach the changing needs of an increasing quota of the population in the more developed countries.

2. TRENDS OF DEVELOPMENT

Some authors see the GNP just as an indicator of the quantitative growth and the increasing GNP, without a qualitative analysis and more sophisticated indicators, appears to them insufficient to detect the whole development.

If 'growth' does not mean automatically 'development', also 'zero-growth' does not mean automatically 'zero-development'. And, in the years 1973-1983, id est in a decade characterized by a GNP zero-growth, many highly industrialized countries developed their socio-economic pattern becoming basically information-and service-economies.[2]

Besides this economic change, in many highly industrialized countries, important social, cultural and demographic developments undergone, during the same years. The main societal developments in the highly industrialized countries in the years 1973-1983, seems to be :

—the decrease of the employment in the industrial factories, mainly the big ones ;

—the increasing employment in the service sector, and especially in the information services ;

---

(1) see : Coen (1978), Deaglio (1982), Etzioni (1968), Galtung and Wemegah (1978), Inglehart (1977) and Thurow (1980).

(2) see : Bell (1976), Ducrocq (1981), Forrester (1980), Kahn (1979) Lipset (1979), Minc (1982)), Naisbitt (1982), Ratenau (1982), Rostow (1983), Simon (1981) and Toffler (1980).

—the increase of the educational level of the population employed in most sectors, and mainly in the information services ;
—the increase of the percentage of the women employed, particularly those with middle and high level of education, engaged in the services and informational sectors ;
—the increasing life's duration in good health, particularly for the women ;
—the decreasing number of children born in an average family, following the birth control and the decreasing infant mortality ;
—the decreasing number of the components of an average family in favour of small families and one-component family ;
—the decreasing attraction of the big cities in favour of the middle towns and the decentralization of the economic activities from the large enterprises to the middle and little enterprises spreading in the periphery ;
—the shift of consumption's peak from the alimentary and basic goods to the pursuit of durable and cultural goods and services.

3. THE CULTURAL LEAD : A SEED FOR FUTURE DEVELOPMENT

The increasing communication allowed by the diffusion of micro electronics at a decreasing price changes the models of reference for the consumption, as well as for the production of goods and services.[3] In the productive side, the available know-how and the institutional and cultural pre-conditions for change are narrowing the gap between the scientific discoveries and their technological diffusion.

This phenomenon, at the national and international level, requires to take in account the potentialities of the scientific discoveries for the economic and social organization of tomorrow.

A foreseeable, rapid and radical change may require a general, strategic planning which is more comprehensive and

(3) see : Brzezinski (1970 and 1981), Dizard (1982), Drucker (1977), Evans (1982) Forrester (1981), Kahn (1979), King (1982 a and b) Littek (1982), Nora (1978), Ratenau (1982), Toffler (1980).

less compulsory than the economic planning by budgeting and requires to foresee some development's scenarios.

The societal change, in highly industrialized countries, may become cumulative and systematic as a result of the behaviour of the mechanisms involved in the change process. Indeed, the differentiation of the life-styles available with an increasing income, offer more room to the 'demand pull' for choosing among rival and/or complementary techniques. Therefore, in highly industrialized countries, the nature of socio-economic change appear to become cumulative, autochtonous and organic more than mechanically induced.

Some authors see a societal change ruled by a technological determinism. I see an area of freedom from such a technological determinism in both : 1) the different ways the same technological innovation may be applied ; 2) as well as in the different times these innovations may spring off and diffuse.

Two main aspects may indicate an area of freedom left besides the technological determinism : the cultural lag and the 'cultural lead'.

The idea that a highly industrialized society may react in different ways to the oncoming technological innovations underlies the polarization of scenarios around a pessimistic approach or an optimistic one.[4] Different time's dimensions and rate of economic growth, besides the alternative ways of difusion, may also explicate the polarization between optimistic and pessimistic scenarios.

4. THE SOCIAL MULTIPLIER EFFECT : A TOOL FOR A STRATEGICALLY PLANNED DEVELOPMENT OF THE HIGHLY INDUSTRIALIZED COUNTRIES

The hypothesis of a 'social multiplier effect' (Coen 1979) consider the starter of a chain of interacting effects which may

(4) it seems to me that some 'pessimist' authors (such as : Jacques Ellul, Lester Brown and Alvin Toffler in his book of 1970 : 'Future Shock') are warrying about a 'cultural lag' failing to keep pace with ongoing changes. On the other hand, I see that some 'optimist' authors (such as: Herman Kahn, Amitiai Etzioni, Dahiel Bell, J.J. Sch reiber and Alvin Toffler in his book : 'The Third Wave' 1980) emphasize a possible 'cultural lead' of the more advanced societies on suitable technobogical changes.

be tested by the long term trends and the strategical social indicators.

A social multiplier effect may be detected, for instance, in the innovative consumption and in the pro-active life-styles which are spreading in the process from invention through innovation to diffusion and translate from one sector to another utilizing some functional interrelations among population's areas and society's subsystems. By analyzing the direction and the time-span taken by the diffusion of the innovative life-styles, some relevant externalities or functional interrelations may be detected.[5]

The inventive, proactive life-styles are challenging both the oligopolistic production and the state-system, i.e. the economic power (acting through the market) and the political power (acting through the political arena). An increasing 'third sector' is related to the diffusion of many inventive or innovative life-styles in more advanced countries. Therefore, the third system is the expanding area for the development of trends of strategical change mutually reinforcing each other.

At a high rate and level of economic growth and urban congestion, a negative feed-back effect is appearing. In the most rich countries, for an increasing quota of the population, the growing GNP does not necessarily reflect incresing longevity of consumer goods, but may reflect increasing production of low-quality commodities and/or increasing social costs or social limits, related to congestion and wastes (see : Giarini and Lauberge, 1978).

Higher education and revenue seem to be related to criticism of material production and congestion. And the results of sophisticated theories of technological impact seem to be empirically perceived by an increasing quota of the population. For instance, in Italy, appears a decline in the population of the big cities and an increase of the middle towns.

These phenomena are of importance for the strategical

(5) see : Attali (1981), Bell (1976), Coen (1979), Colombo and Turani (1982), Etzioni (1968), Frreman (1983), Gershuny (1978), Giarini and Lauberge (1978), Gorz (1980), Lipset (1979), Minc (1982).

planning[6] because they show which of the societal choices, besides the induced behaviour, are the results of the state's or the market's choices. Many societal preferences for the future appear in the behaviour of the leader groups, (mainly the most informed and young people of the post-industrial societies) because the increasing information inducing criticism, joines the increasing 'discretionary income' allowing the freedom to choose how better to use the income above the necessary one to fulfil the basic needs.

The diversification among consumption styles and the widening of social behaviours seems to move these 'proactive' rather than reactive live-styles from a first stage (supercompensation of a recently overcome deficiency of consumption) to a stage of consumer proactive choice seeking for individual self-expansion.

In order to allow personal self-expansion, both the state and the mass-production cannot provide individualized services with efficiency. Therefore, there is a need for a third sector having the necessary flexibility to face the changing individual needs. The intangible goods and services are hardly measured by economic data and statistics and therefore impossible to manage either by a system or by a mass production market.

---

(6) The type of strategical, longterm planning I am speaking about is a methodology of planning which uses the societal mechanism a declared purpose. Such a methodology is grounded on knowledge more than on power and is focusing on study and research more than on administrative roles. It is a soft methodology of planning, not a hard one, because it will lead the society by using the existing mechanisms instead of imposing on the living social system a heavy economic or bureaucratic structure.

This strategical planning methodology, mainly based on the knowledge of the societal system, will utilize a political methodology soft in its means but revolutionary in its longterm and large-range results, because the strategical planning is a methodology speeding up some societal mechanisms in order to reach long-term and large-range effects.

For a strategic, future-oriented planning, also the communication media and the information contents appear to become of strategical importance for their social multiplier effect. Both Daniel Bell and Amitial Etzioni focused on the knowledge's property to diffuse from one to another increasing the general output.

The intangible are of difficult mangeability for big organizations and their treatment may become dangerous, because public intervention in intangibles may increase the risk of people becoming prisoner of benevolence of public parenthood". This danger is growing when public intervention is not self-restraining to the "pure collective" goods and services.

Public intervention for satisfying social needs may increase the State's role as a supplier of goods and services, with a consequent expansion of bureaucratic administration and a consequent concentration of power in the political arena and a massification of society. Private intervention, on the other side, focusing the satisfaction of social needs may increase the market's role with consequent concentration of power on personal and family wealth and an increasing social differentiation and inequality.

A new direction, between the State and the market, as suggested by Ruffolo (1980), may increase the role of a third sector of organizations and agents of change that are growing where a critical consciousness is diffusing among people. Such a third system may return to the people an autonomous power, reducing the influence of the first and the second system (respectively the State and the market) over the destiny of people who struggle for more freedom and diversification, by constituting new movements, free associations, groups of militants for participative and cultural activities, voluntary social work, etc.

REFERENCES

ATTALI J. (1981) : '*Les Trois Mondes, Pour une theorie de l 'aprcs-crise*' Paris, Fayard,.

BELD D. (1976) : '*The Coming of Post-Industrial Society. A Venture in Social Forecasting*' Basic Books, New York.

BRZEZINSKI, Z. (1970) : '*Between Two Ages : America's Role in Technetronic Era*' ? Viking Press, New York.

BRZEZINSKI, Z. (1981) : '*The Technetronic Revolution*', Viking Press, New York.

COEN A. (1978) : *Innovative Processes in the Life Styles of Industrialized*

Countries'. Paper discussed at the IX World Congress of the International Sociological Association, Upsala, Sweden, August.

COEN A. (1979) : '*A Direction for Long Term Development* : *Quality versus Quantitv*'. Paper discussed at the World Futures Studies Confcrence, Berlin, May, 1979 and published in : Bucholz-Guelin (ed) : '*Science, Technology and the Future*' Sauer Pub. Munchen and New York, 1979 ; pp. 396-414.

COEN A. (1980) : '*Influenze del Nuovo Ordine Economic Internazionole sulla funzione sociale c redistributive a sullo stile di vita e la valozzazione del capitale umanonci paesi della Comunita Europea*' pp. 233-248 di '*Affari Sociali Internazionali*' anno VIII—n. 1-2, 1980.

COLOMBO U. TURANI G. (1982) : '*Il Secondo Pianeta*'—*Mondadori ed.* Milano.

COLOMBO U.—LANZAVECHIA G. (1982) : '*The Transition to an Information Society* : *How Do we Manage to Change* ? Wat can RD Contribute ? pp. 3-35 in A. V. : '*Information Society*' *North*, Holland, Amsterdarm.

DEAGLIO M. (1982) : '*Come cambia il Capitalismo*'—Mondadori ed, Milano.

DIZARD W. P. (1982) : 'The Coming Information Age : an Overview of Technology, Economics and Politics', Longman Inc. New York.

DUCROCO A. (1981) : '*Vers une Societe de Communication*' *Hachette*, Paris.

DRUCKER P. F. (1977) : '*The Innovative Organization*, pp. 3-13 in Dialogue vol. 101.

ETZIONI A. (1968) ; 'The Active Society', Free Press.

EVANS J. (1982) : '*The Worker and the Workplace*' pp. 157-182 in : Friedrichs, G. and Schaff, A. : Microelectronics and Society' Pergamon Press, Amsterdam.

FORRESTER J. (1981) : '*Innovation and Economic Change*' in '*Future*. n. 13.

FORESTER T. (ed.) (1980) : '*The Microelectronics Revolution*' Blackwell, Oxford.

FREEMAN C. (1982) : '*Unemployment and Technical Innovation* Pinter.

FREEMAN T. (ed.) (1983) i '*Long Waves in the World Economy*, Butter worth Pu. England.

GALTUNG J. and OEMEGAH, M. (1978) : '*Overdevelopment and Alternative Ways of Life in High Income Countries*', United Nations University Tokio.

GERSHUNY J. (1978) : '*After Industrial Society* ? *The Emerging Self-Service Economy*'. Macmillan Press, London.

GERSHUNY J. and MILES J. (1978) ; '*The New Service Economy* : *The Transformation of Employment in Industrial Societies*, Pinter Pub. London.

GIARINI O.—LAUBERGE' H. (1978) : 'La delusione technologica' Mondadori, Milano.

GORZ A. (1978) : 'Adieux au Proletariat' Galilee, Paris.

HAMRIM R. D. (1981) : 'The Information Economy' : an Infinite Resource' in 'The Futurist', August.

HIRSCH F. (1978) : 'Social Limits to Growth' A Harvard Paperback, Cambridge Mass.

INGLEHART R. (1977) : 'The Silent Revolution' Princeton Univ. Press, Princetow New Jersey.

KAHN H. (1979) : 'The Coming Transition', New York.

KING A. (1982) : '*A New Industrial Revolution or Just Another Technology* ? Introduction to : *Friedrichs*, G. and Schaff A. : *Microelectronics and Society*, Pergamon Press, Amsterdam.

KING A. (1982) : '*For Better and for Worse. The Benefits and Risks of Information Technology*'—pp. 3-56 in A. V. : 'Information Society' North Holland.

KUTSCHER R. (1981) : '*New Economic Projections through 1990 : an Overview*' in '*Monthly Labor Review*', August.

LAWRENCE R. Z. (1983) : '*Is Trade De-Industrializing the United States* ? In '*Brookings Papers on Economic Activity* p. 1.

LIPSET, S. (ed.) (1979) : '*The Third Century* : *America ad a Post-Industrial Society*', Institution Press, Stanford University Califoinia.

LITTEK W. (1982) : 'New Technologies and the Development of Qualifieations' Paper to the 10th Congress of Sociology Mevico City, August.

MINC A. (1982) : '*L' Apres-Crise est commence* Gallimard Paris.

MOMIGLIANO F. (1983) : '*Problemi di analisi delle determinanti ed effetti delle innovazioni nei paesi industrialli avanzati e in Italia*', Relazione presentata al Convegno di Nomisma, Milano, dicembre.

COCIGLICNO F. (1983) : '*Determinanti ed effetti delle spese di* RS in una industria ad alta Opportunita Tecnologica' in '*L' Industria*' *n. 1*, Il Mulino ed. Bologna.

MORIN E. (1981) : '*Pour Sortir du Vingtieme Siecle Dossier Fernand Nathan*, Paris.

NAISBITT J. (1982) ; '*Megatrends*' Warner Pub.

NORA S. (1978) : '*L' informatisation de la Societe*' *Editions du Seuil*, Paris.

OCDE (1975) : '*Ajustement en vue des echanges.* Etudes dee problemes et politiques d'ajustement industriel' OCDE, Paris.

OCDE (1976) : '*Politiques visant a promovoir I 'adaptation industrielle*' OCDE, Paris.

OCDE-INTERFUTURES (1979) : '*Final Report Facing the Future*', OCDE.

OCDE (1980) : '*Technical Change and Economic Policy*, OCDE, Paris.

OCDE (1983) : '*L' industrie en Mutation*' OCDE, Paris.

OLSON' W. (1982) : '*Scelta Pubblica e Sviluppo*' pp. 87-100 in : *Economia Pubblica*' *anno 12* n. 3 : marzo.

OLSON M. (1982) : '*The Rise and Decline of Nations* : *Economic Growth Stagflation and Social Rigidities*' Yale Univ. Press., New Haven.

PINDER J. (ed.) (1983) : '*National Industrial Strategics and the World Economy*' *Allanhell*. New Jersey.

RATENAU G. W. (1982) : '*The Impact of Information Technology on Social Organization and Structure* pp. 57-66 in A. V. '*Information Society*' North Holland, Amsterdam.

ROSTOW W. W. (1983) : '*The Fifth Upswing and the Forth Industrial Revolution*' in *Challenge*.

RUFFOLO G. (1980) : '*Complessita e Democrazia*' : *il Ruolo della Informazione in* 'Critica Marxista'.

SIMON J. L. (1981) : '*The Ultimate Resource*', Princeton University

THUROW L. C. (1980) : '*The Zero-Sum Society*', Basic Books.

TOFFLER A. (1980) : '*The Third Wave*', Collins, London.

U.S.A.-THE GLOBAL 2000 (1979) : 'Report to the President, Washington.

WESCOTT R. F. (1983) : '*U.S. Industrial Approaches* : *a Review*' *in Adams F.G. and Klein L. R. Industrial Policies for Growth and Competitiveness*. Lexington Books. 1983 c sul p. 45 di 'Economic Impact', 1.

10

Towards the Digital Society ?

Victor Scardigii*

How far have we got, on the way to a digital society ?

I am referring to this term, because of the less precise definition of the 'information society' (is it a society which functions according to the laws of the information theory ? Is it a post-industrial society, reaching a given threshold of workers involved in various fields of information processing? Or, more widely, a society where information becomes the most valuable good, and where all goods become informations?[1]).

Technical advances in micro-processing, electronics and tele-communication offer a large range of goods and services, which are investing more and more sectors in modern societies, more and more fields in our daily life ; the term 'digital society' focuses on the long-term influence of these tools on culture and way of life.

According to the prevailing predictions in the late seventies, these digital tools were to give shape to any individual's activity, to the whole space-time framework of the daily life, and more deeply to our representations and value systems. North-

1. This analysis is partly referring to P.A. Mercier, F. Plassard, V. Scardigli, La societe digitale. Les nouvelles technologies au futur quotidien. Le Seuil, Paris, 1984.

* Research Fellow at CNRS, Centre National De La Recherche Scientifique, Paris.

American and Japanese forecasts (Toffler, Masuda...) reveal a strong confidence in the triumph of technical advances over any major problem in economies and societies. European ones (Nora—Minc) were more prudent regarding the pace of technical change, and more pessimistic on the nuisances of social change ; but there was no doubt on the output : digital tools would necessarily produce a digital society.

These first attempts were akin to technological forecasting methodology : starting from new goods and services which were feasible, they guessed their possible uses, and then their second-range effects on society. Nowadays, still, American forecasts often go on using this method : they seem to be confident in the preponderant strength of the technological wave.

This is why we (two French teams in the Centre National de la Recherche Scientifique, in connection with the European Commission F.A.S.T. Programme) have tried to explore an alternative route for researches on the future of digital societies.

To my opinion, two methodological principles can lead to such a route : we have to be more concrete, observing how the diversity of interactions society, technologies does actually develop ; and, altogether, we have to be more theoretical, connecting these developments to the broader framework of the technopolistic society.

MASTERED INTEGRATION, SUBJECTION, DYSFUNCTIONING : THREE POSSIBLE INTERACTIONS BETWEEN TECHNOLOGY AND SOCIETY

In previous works, future developments were often considered as if they were engraved in the very core of the technology, like a genetic programme ; the task of the sociologist being to decode this destiny.

Yet, more and more empirical data become available, showing that the actual social behaviours related to the new information and communication techniques are of a large diversity, and, to a large extent, quite different from the inventors', manufacturers' and States' provisions. The video was introduced as a means for a renewal of conviviality (small groups doing their own information films on local daily life, problems at a local community level), and for personal creativity (each individual creating his own frames and sounds); it is actually used in a rather passive way : the V.C.R. is just like a supplementary TV channel, casting commerical films. On the reverse, the telephone was sold for a passive reception : music and bel canto were to be played in a central place and cast to private homes throw the wires : the inter-active use for communication between individuals was unforeseen.

Thus, technological advances cannot be studied apart from society : the social dynamics is permanently defining a social demand for new tools and their social uses, as well as the technical offer is shaping the way of life (considered in the sense of the coherence of daily individual activities, of values and representations, specific for a given social group of culture, at a given stage of its history).

It becomes essential to gather field observations on this inter-action between tools and men : at least, on the first uses and users in education, trade, out-of-work daily life (computer assisted instruction, electronic transfer of funds, micro-computers, video-camera and recorder, data banks,...) ; in the work-place (computer assisted design and manufacturing, office automation,...) ; in so-called social experiments on community networks, and so on.

These observations, made in France, but altogether in several countries of Europe, North America and Third World, lead to conclusions which often are contradictory, at least extremely diverse, on the ways of men-to-tool and tool-to-men adaptation, of individual and collective appropriation, of cultural upheavals... This diversity is meaningful, and I shall schematize it according to three 'interaction models'.

In the first model, of a mastered integration, there is a convergence of social needs and of technologies: changes in the way of life call for new goods and services, which in turn are easily adopted and, as a feed-back, help people producing a tomorrow society which answers their aspirations.

In model two, of subjection, supply tries to create demand, and the daily life is invaded by the new techniques. People cannot avoid using them for their survival, but they get controlled, they develop dependence and passive attitudes.

In model three, of dysfunctioning, a divorce occurs between

a supply dictated by producers and by the State, and a population who rejects this graft, struggles for the survival of a traditional non-technical way of life, or leads the new tools astray from their official uses, finding out unforeseen applications.

Mastered Integration

In this model, innovation develops in response to needs of individuals who integrate harmoniously information and micro-electronic techniques in their daily lives. Capable of using to their advantage the possibilities offered by modern technology, people actively see progress automated work and market consumption as a means of assuring the modern relationship of the individual with the world, with others and with himself.

To this behaviour corresponds, on the supply side, a very fast and all-directional distribution of electronic and computerized goods and services, both in the form of network and non-connected froms. All these techniques produce information which is abundant and of quality, responding well to the needs of households and firms. Programmes are simple to learn and to use : they avoid all errors, they incorporate many diversified skills and take account of national and regional peculiarities.

Daily life in this model is characterised by a depreciation of domestic duties, the return to a new sociability and the fastened emergence of the individual.

Certain duties, characteristic of traditional society (preparation of meals, wash-up, child care), are endured as chores, drudgery, because they have no economic value and they do not incorporate scientific advance. To carry out domestic tasks, the individual uses all the resources offered by a new technology ; e.g. for planning the variety and contents of meals or for memorizing family information and events. They have wide access to the possibilities of T.V. and video to supervise children and to practice tele-shopping on a large scale.

Far from aggravating the damage of industrialisation and of urbanisation, technology is used to compensate for this damage and to recreate a web of sociability and of solidarity. In so far as new techniques permit a real conversational use, tele-shopping does not eliminate local commerce. In the same way, the telephone and peri-telephone products allow faraway relative or friend ties to be maintained and keep local associations actively alive. Instead of watching T.V. during home meals and conversations, people record T.V. broadcasts while preserving relationships inside the family.

Each individual behave as a basic unit of society. All the equipment and new services give him the mobility and the autonomy required by his new roles. Tele-information and reservations from home or from a shopping-center in the neighbourhood, credit cards and electronic transfer of funds permit the individual to save time and especially to organise his use of daily time, while forgetting the constraints of normal opening hours. Telework in workshops near home allows a minority of individuals to alleviate the break between work life and outside work life, as does the consultation of data bases from home.

Through all these activities, the individual finds happiness and his personal fulfilment. But in this search he keeps the initiative of the activity that he wants to participate to, he knows where and how to find the information that will be necessary to him. This attitude extends the productivist ethic adopted by the industrial mankind : happiness and fulfilment of the individual are certainly the ultimate goal of daily life, but work, the accumulation of goods and activities, remain the means of achieving this objective.

Subjection

This second model sees the domination of supply over demand. Manufacturers succeed in dictating necessary uses of the new techniques in work and daily consumption : population is transformed into uprooted, helpless, powerless individuals, who accept passively this imposed modernity.

Just like the worker in the Taylor revolution, sometimes the individual is deprived by the machine from his traditional competences : cooking, counting, leading administering, playing, learning. Excessive urbanisation, the destruction of networks of traditional solidarity find expression in a new reliance by the individual on his home, invested furthermore with its numerous goods and services. These develop to the detriment of traditional cultures, which are at the same time crushed by standardisation and the universalisation of programmes, and destroyed by the transfer of skills from man to machine.

The non-market sectors of daily life are progressively integrated into the area of market consumption, leading to the reduction of the informal sector and of actual social life.

New technologies carry poor content and are reproductive of previous daily activities, instead of generating new patterns behaviour.

Information technologies are mainly sold in the sectors of consumption actually underway : leisure activity and household comfort in particular. In the area of leisure activity, they are used in their dimension of automation, of facilitation, of improved reproduction and not as the instrument of new possibilities of creation and personalisation. The computerising of domestic activities is hardly visible, so as not to frighten the user, who has not, due to this, any possibility of understanding the functioning logic of the apparatus. Their programme corresponds to an increased standardisation which constrains the individual to adapt to the new mode of consumption. Their is certainly facilitation of tasks and a gain of time, but this is most often absorbed by passive consumption of new mediated leisure activities.

This model, therefore, is characterised by a gap between the increased possibilities of consumption of new goods and services, and a low level of technical knowledge which does not allow individuals to master the terms and meaning of these new consumptions. It implies a relation of subjection, of incomprehension and of passivity toward new technology.

Dysfunctioning

In this third model, culture and technology from two worlds, separated by the refusal of individuals to rationalize daily activities and out-of-work time. Electronics and data processing do not meet actual needs, they remain gadgets without influence on social life.

Lacking demand, the new technology begins with applications depending on the State : for example, the French decisions for a compulsory electronic directory, and for the nation-wide cable-laying programme. Networks, essentially centralised and specialised, are not always connectible, due to insufficient standardisation. Programmes remain complex and abstract. On the screen, alpha-numeric data dominate over images.

Daily life is only slightly based on the new services. Home videotex is hardly used for long-distance buying, nor for obtaining practical information. In several cases, people invent unforeseen or illegal uses, for instance individual messages, love affair parallel network... Young (or older) pirates find free access to big computers, they do their own data processing, or transfer money to their own banking account, or reveal secret information. VCR piracy develops. People use micro computers just like electronic games, ignoring their high processing potentialities.

Stored information is of uneasy access, procedures are rigid and without connection with the usual ways of thinking.

Fear of computer filing and of identity centralised systems leads to deviant behaviours : some people do not heed recommendations given by the system, others happily and willingly provide false information. Workers find how to bring office automation and c.a.m. to failures. Although actively supported by public authorities, eomputer assisted instruction has to face teachers' unwillingness and shortage in available recorded programmes ; a poor adaptation of programmes to non-codable matters causes a lowering of standards of pupils and an accentuation of inequalities.

People's lack of confidence in the machine and always need to check-up any information given by networks.

Suspicion for new technology sometimes lead activist groups to blot data out of computer memories, or even to destroy big computers.

The push toward rationalizing daily informal activities gives rise to a hostile attitude which reflects a fear of losing control of one's life and environment, a will to protect family intimacy from electronic invasion, a priority given to traditional uses and values. A large sector, non-computerised and obeying a non-market logic develops parallel to the official economy.

Leisure develops on the fringe of the technique ; the search for a non-media guided conviviality leads to a revival of neibourhood associations and local community life.

As a whole, this third model is a dysfunctioning one according to a technical and economical rationality ; but it may prove interesting from a sociological point of view. Of course, if there are too many robberies and voluntary breakdowns, if people experiments too many hardware failures and software problems if they refuse the digital environment or try to get rid of it, States and firms will fear an economical failure and slacken the pace of innovation, the digital revolution will abort.

But, when the new tools are used in unforeseen ways—invention of new uses, misappropriation, passing of programme...then these behaviours can lead to new forms of mastering techniques ; thus, they are social innovation, even more than in the first model.

These three models of interaction are not exclusive : they may sometimes be observed for the same type of new technique used by different countries or groups, or even for the same person using different goods and services. This diversity prevents from trying to give to-day a final response to interrogations about the future of digital societies : things are not yet played, products are not settled nor their uses, several futures are possible.[1] One more conclusion emerges from this first part of my paper. The uncertainties must not hide the fact that renewed difficulties are preparing for underprivileged groups and poorer countries. It comes to evidence that the mastered integration model mainly characterizes people (upper and middle classes, young, male rather than female...) who have few financial problems about the cost of

1. Another part of our research has studied thoronghly four global scenarie. Each of them combines the previous models and contrasted frameworks of economic environment, State intervention or lazssel-faire, and social innovativeness : crisis wtthout social change, adapting crisis thanks to new technologies, growth thanks to information goods, information soeiety thanks to growth.

access, to networks, data banks, new products, etc., and even fewer educational problems about learning how to make use and reap advantage of the new tools. More widely, it corresponds to big firms and dominant States which are promoting these new technologies.

On the contrary, subjection is the extension of a situation of inferiority and dependency which is already well-known in the cases of pupils from a proletarian origin, of low-skilled workers, of old people facing any new technical advance, etc., In the same way, the irruption of new technologies in the Third World nations is far from giving them a "technical short cut' for development ; it often proves increasing their dependancy on software, data banks, TV series,..., imported from economically and technically dominant countries. In a more covert way, the South countries import altogether a way of life, a values system, an economic rationality which belong to western culture : they are destroying the roots of their own traditional cultures.

STAGES IN THE DEVELOPMENT OF A DIGITAL/ TECHNOPOLISTIC SOCIETY

Were our analysis only leading to this typology, it might give the impression of a timeless and changeless coexistence between three technical/economic sectors, three social/regional groups...

Now, the situation has much changed for ten years, and several stages can be discerned in the development of the digital universe. This calls for a more theoretical framework of explanation: let us briefly describe what these stages are, before trying to relate this historical dynamics to other previous examples of diffusion of modern goods and services in our type of society.

At the beginning, during the late seventies, the new technologies had little diffusion : one could hardly guess what they would be like, in several cases. Therefore, it was what I could call the period for phantasms and manoeuvres.

It was the stage of technological triumphalism, leading to great hopes and fears.

The social changes that were foreseen, linked to information technologies, belong to a set of seven hopes which all converged toward "humanity happiness through scientific and technological advances". At the opposite, some anxious positions seen in public opinion polls as well as in several decisions evoked an opposite set of fears : the regret for a lost Golden Age.

This mythology of technique had seven bi-polar dimensions :

—man's liberation from any boring task at home or at work rediscovered autonomy in use of time for personal training, self-fulfilment, leisure.../versus : robot which makes man a slave, T.V. passivity...;

—access to the whole human knowledge/versus : children can no longer read, write and count ; loss of 'technical culture' ;

—decentralisation, or even coming back to nature, from where you can work, buy, receive information, get in touch with everybody/versus ; destruction of informal social life ; technologies rise as screens between men ; their binary logic eliminates all that is not rational, destroys any affective communication ; increased control of the States and of Multinational Corporations on our daily life ;

—answer to economic problems (little energy needed, no pollution ; and the demand for these new goods will stimulate growth)/versus : crisis aggravation (unemployment, technological delay of Europe) ;

—thanks to a low-cost access to any knowledge of mankind, decreasing inequalities among developed countries/versus : the rich and well-educated people will be the main users of data banks or training facilities ;

—technological short-cut for Third World/versus : increased gap between North and South countries.

We can find again this collective imagery in official statements of this period, as well as in advertising, mass media, novels of the late seventies.

Nobody knowing which of these forecasts would be confirmed, hopes and fears gave support, at the same time, to great manoeuvres. Big firms created transnational data banks and data processing networks. They convinced populations that the economic survival would require the creation of a large inner market for the new goods and services; the so-called communication tools were to bring their magic solution to the increasing problem of lack for social communication in modern societies ; each citizen would tomorrow have to use data processing as currently as reading and writing. And the governments took great decisions regarding the cable and home telematics, computer at school, etc.

Then came the second stage: enthusiasm fell down, passions abated. One could see the first outcome of office automation and computer assisted work : an increasing unemployment, as least in Europe. If pessimistic prophecies often were right in the field of work, costs and benefits were less evident in most fields, and governments became—and are still—hesitating : cable or satellite ? local or national TV networks ? collective services, or private consumption ? And they delay or cancel their previous decisions. Electronic games seem to reach a sales ceiling, financial problems upset the personal computer industries, the Japanese fifth computer generation seems to be but a bluff.

In out-of-work daily life, electronics has known a rather fast diffusion : V.C.R., cable TV, micro-computers, payment cards, 'hidden' electronics in domestic robots or cars, etc, are spreading. But their first uses reveal reproductive (the same kind of leisure...)instead of being innovative. As a whole, the technical advance does not seem to upset anything in culture and way of life.

Then rises an impression that, according to La Fontaine's fable, the muntain gives birth to a mouse. The digital revolution brings only new gadgets, it has no major social impact.

Must we stop here our analysis ? A sociologist's role is to pay attention to covert changes which are diffuse, scarcely visible to-day. Even if information and communication tools seem often to be mere gadgets, they are weaving a network which encloses each of us more and more tightly. Such a quantitative accumulation of technical interfaces between man and his physical and human surroundings leads to qualitative changes : in a third stage, their long-range impact may reveal crucial for man and society.

This is a lesson brought by a much larger range of modern goods and services which all combine *tekhne* (high competence of producers, integrated in the consumption goods) and *polein* (to have an individually or collectively payed access to need satisfaction through these products). The development of a digital society reveals to be only a facet of larger phenomenon: the coming of a technopolistic society.

Let us take two examples, chosen in opposite fields of innovation, of social and economic dynamics, and of unnecessary/fundamental needs : the diffusion of automobile cars, and of modern medicine, have known the same three stages.

In a first stage, car was almost a fantasm : a symbol for liberty and personal fulfilment (it meant spatial mobility, escape from town and work, access to higher status, power and virility, enjoyment...), as well as a threat of class inequality, individualism, waste... Biological advances were abolishing pain and promising immortality ; the end of natural selection, genetic manipulation were leading to 'Brave New World' phantasms, like Orwell's '1984' in our electronic field.

Leaning on these hopes and fears, we saw the expansion of these economic sectors which became 'growth poles' ; pressure groups settled gear-wheels which avoided our economy and ways of life from going backward to traditional definition and answer for needs in these fields.

Then, on the surface, came a second stage, which rendered these consumptions commonplace. Social differences seemed to reduce, one thought that a car was merely useful, that health just a basic need.

But a further analysis[2] shows that major changes were preparing for a third stage. Automobile has changed our relationship to our physical environment : space is perceived as a boundless network, time is varying according to the means of transportation. Car has contributed to a scattered urbanization,

2. V. Scardigli. La consommation eulture du quotidien, Paris, Presses Universitares de France, 1938.

to crumbling daily activities of any person. It has begun denying the social group and promoting the individual as the basic cell of the to-morrow society.

Modern medicine has changed our conception of life and body. No longer of a metaphysical origin, the body is a physical machine that you can fix, or even modify in order to make it consistent with a person's desires, and with society's expectations for performance. Thus, medical advance is developing the technopolistic process : man is deprived of his body and his destiny by experts ; one can pay for setting up a medical draught-screen which hides fear of death and culpability.

Let us notice the steps forward : in the first case, the transformation remained mainly outside, even if Piaget has shown that the child is modified by his relationship to the outside space. In the second case, medicine begins hitting the very man, the behaviours he has ; but, mainly, his body.

In the case of the information and communication technologies, the transformation goes still further. In a very long term, there will probably be a change in location of economic activities and exchanges, educational institutions, leisure, etc., But, here and now, we can see a modification in the whole cognitive and affective relationship to oneself and to the other man.

Any knowledge is quantified, measured, payed. Any cognitive process calls for experts who help perceiving (robots for vision and voice recognition), or even thinking, deciding (expert systems and artificial intelligence) ...More widely, the work-universe rationality is invading out-of-work life. Free time is no longer free, it is over-occupied by the paying consumption of audiovisual media, and means of communication, of plays, of creativity with sounds and colours. The contact with the other is measured in terms of time, cost, interest. All problems of the planet (starvations, slaughters) becoming omnipresent, this may lead to indifference toward this abstract-visible mankind.

More basically, the very relationship to reality is upset by the new technologies. Advances are multiplying the sophistication of our daily tools, in which we are less and less able to understand an increasing number of 'black boxes'. The conversion of any knowledge and communication, of the whole reality into digits results in a situation where things have less and less link with actual sounds, images, beings. Digital prostheses and screens are restoring a reality where digital simulacrum looks more real and perfect than nature. After having invested the physical environment, and then the human body, the process is going to extend now to man's knowledge and heart, to society's communicational life, to reality. In this stage, mankind will live in a technopolistic future.

Index